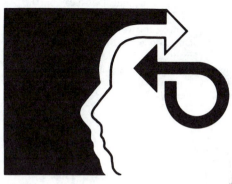

**SERIES IN
PSYCHOSOCIAL
EPIDEMIOLOGY**

VOLUME 8

GENETIC ISSUES IN PSYCHOSOCIAL EPIDEMIOLOGY

EDITED BY
Ming T. Tsuang
Kenneth S. Kendler
Michael J. Lyons

 Rutgers University Press
New Brunswick, New Jersey

The editors would like to recognize the support of grants DA04604 from the National Institute on Drug Abuse, R37-MH43518, MH44277, and UO1-MH46318 from the National Institute of Mental Health, and a Veterans Affairs Medical Research Merit Review Grant to Dr. Tsuang and grants MH40828 and MH41953 from the National Institute of Mental Health to Dr. Kendler that helped to make this work possible. We would also like to recognize the valuable assistance of Mary E. Merla and Jennifer E. Boyd to the successful completion of the myriad tasks associated with producing an edited volume.

Library of Congress Cataloging-in-Publication Data

Genetic issues in psychosocial epidemiology/ edited by Ming T. Tsuang,
  Kenneth S. Kendler, Michael J. Lyons.
      p.   cm. —(Series in psychosocial epidemiology : v. 8)
    Includes bibliographical references.
    Includes index.
    ISBN 0-8135-1729-X
    1. Mental illness—Genetic aspects.   2. Mental illness—
Epidemiology.   3. Psychiatric epidemiology.   I. Tsuang, Ming T.,
1931–   .   II. Kendler, Kenneth S., 1950–   .   III. Lyons, Michael
J., 1949–   .   IV. Series.
    [DNLM: 1. Epidemiologic Methods.   2. Mental Disorders—
epidemiology.   3. Mental Disorders—genetics.   W1 SE718R v. 8 / WM
100 G3285]
    RC455.4.G4G456   1991
    616.89'042—dc20
    DNLM/DLC
    for Library of Congress   91-11720
                                                                         CIP

British Cataloging-in-Publication information available

Dedicated to

Snow, John, Debby , and Grace (M.T.T)
Susan, Jennifer, Seth, and Nathan (K.S.K.)
Kathleen, Morgan, Drew, and Ashley (M.J.L.)

Contents

Introduction to Studies in Psychosocial Epidemiology

Rising costs of mental health care and interest in the development of national health insurance that should include some mental health services have heightened concern both for ways by which health planners may evaluate the extent of behavioral problems and for means by which the efficacy of various treatment modalities may be assessed. Questions of paramount importance include: If two therapeutic modalities are equally effective, which is more efficient? If equally efficient, which is less costly? Comparable questions regarding the prevalence and incidence rates of specific psychiatric illness in communities are being raised by those delegated the task of making policy recommendations for programs of primary prevention and treatment. What is the natural history of an untreated behavioral problem and how is the problem affected by normal growth and development? What does knowledge of the natural history of a disorder of mood, thought, and behavior tell us about its etiology, and how may such knowledge lead to methods of prevention? If an illness is not preventable by currently available knowledge, what interventions may be made in its natural history to arrest and possibly reverse its course? Mental health is now big business; failure to look at specific population needs when planning programs and to include ways of evaluating cost, efficiency, and effectiveness results in considerable psychological and economic cost to millions of patients and their families, as well as to taxpayers in general.

Epidemiology is a body of knowledge and technical skills that may be put to use in answering some of the questions facing health planners today. Traditionally, epidemiology has been seen as the study of disease patterns in populations. Epidemiologists have been useful in providing data that have led to effective programs of prevention of a number of infectious diseases including malaria, smallpox, and poliomyelitis. Epidemiology has, however, played a minor role in psychiatric research until relatively recently. Epidemiological studies in mental health tended to be descriptive and to focus on the prevalence and incidence rates of symptoms in broad categories of illness, such as "neuroses" or "psychoses." Some infectious- and chronic-disease epidemiologists, in fact, question whether epidemiology can be used to tackle psychiatric problems. The Society for Epidemiologic Research does not have a section on psychosocial epidemiology, and publication of papers in social and psychiatric epidemiology in its journal is infrequent. Principal organs of dispersion of knowledge in psychosocial epidemiology have been *Psychological Medicine* and the *Archives of General Psychiatry*. The former journal, published quarterly has on its editorial board members sophisticated in epidemiology and probably publishes the greatest number of articles in this area. It, however, is a British journal with limited readership in the United States. On the other hand, the *Archives of General Psychiatry*, a publication of the American Medical Association, is fairly widely read in the United States. It publishes a number of high quality papers in epidemiology but has a broad mandate and, therefore, is limited in its ability to publish more.

The Series in Psychosocial Epidemiology, of which this is the eighth volume, is to serve important functions for epidemiology, psychiatry, and related areas in public health. Its objectives include:

1. Providing a forum for discussion of research strategies in the evaluation of mental health problems in the community and for assessing the effectiveness of psychiatric treatment interventions,
2. Serving as a teaching tool for students in medical schools and schools of public health, hospital administration, social work, and nursing, as well as for students in departments of psychology, sociology, and especially epidemiology,
3. Keeping prospective researchers alert to problems needing investigation in the area of psychosocial epidemiology,
4. Serving as a vehicle to bring together research in the area of psychosocial epidemiology,
5. Providing a means of continuing education for epidemiologists working in the field,
6. Providing a means for discussion of how the results of epidemiological and related behavioral research might be

brought to bear on the development of state and federal health policy decisions.

To achieve these ends, each volume has a single theme of particular interest to investigators in the field of psychosocial epidemiology, themes such as the study of children, the study of stressful life events, and needs assessment. Each volume in the series has a guest editor or editors, with established reputations in the field who have selected and edited contributions and written the introductory article. In general, in the lead article the guest editors (1) discuss methodological considerations for research in a given area (i.e., experimental design, sampling, instruments used, analysis of data, economic considerations, and ethical decisions in planning studies), (2) critically review existing studies, (3) present information on the current state of the field, and (4) suggest directions for further research. I discuss in the Foreword the present state of epidemiological research in the area, the directions the Center for Epidemiologic Studies at the National Institute of Mental Health would like to see research take, and the relationship of research in the field to the formulation of health policy.

Earlier volumes in the series focused on studies of children, stressful life events, help-seeking behavior, alcohol use studies, drug abuse studies, community surveys, and ethical issues in epidemological research. The mandate, however, of this series is challenging and the task great. With the help of the guest editors and other contributors I am sure the series can provide researchers, students, and health policy makers much of what they need to know about research in psychosocial epidemiology in order to plan future research and to design effective and consumer-responsive health policy and preventive psychiatric care problems.

A.E.S.

Foreword

This, the eighth volume in the Series in Psychosocial epidemiology, is a testament to the rapidity of advances in the understanding of psychiatric disorder using modern epidemiologic techniques. The senior volume editor, Ming T. Tsuang, Professor and Vice Chairman for Research at Harvard Medical School, is an internationally recognized leader in the establishment of psychiatric genetic epidemiology as a subspecialty of psychiatric epidemiology. He shows his personal breadth of understanding of both the subject and the methodology by the contributors and topics he and his coeditors, Kenneth Kendler and Michael Lyons, have chosen for this volume. The material presented is of major concern for investigative scholars in genetic epidemiology, as well as for clinicians and concerned health planners interested in genetic counseling, family planning, and preventive medicine.

As a review of the literature supporting the various chapters indicates, significant advances have occurred in the actual months preceding publication of this volume that have further elucidated the role of genetics in the transmission of mental disorders.

Traditionally, understanding of familial factors in psychiatry was derived from case registries, family histories, twin studies, and adopted-away studies. These all were limited by the level of ability to diagnose a disorder and understand the phenomenology of variants of the illness, by professional and societal taboos, and by the level of skill in eliciting family and individual histories. Reported current increases in some disorders and diminution in frequency of others may represent not a real change in incidence or prevalence but rather advances in diagnostic techniques such as those discussed in an earlier volume in this series (edited by Myrna Weissman and her associates, who are employed in Epidemiologic Catchment Area studies supported by the National Institute of Mental Health) and increased willingness of affected individuals to discuss their own symptoms, past histories, and family histories. Studies in molecular biology complement those of genetic epidemiology using linkage analyses and other advances in statistical understanding to direct future research and enhance further understanding of genetic transmission.

Obviously all of us, whether scientists or concerned lay people, are entitled to be kept abreast of how disorders such as bipolar illness, major depression, criminality, schizophrenia, alcoholism, other substance abuse disorders, anxiety disorder, obsessive disorders, and impulsivity resulting in suicide or homicide may be transmitted. Clearly, such an understanding must be presented in an ethical and critical manner given the distortions and misuse of data that have occurred both medically and politically. Family planners and genetic counselors require the state-of-the-art understanding presented here to provide their clients with answers to questions regarding the adoption of children with known family histories of psychiatric illness, as well as to counsel couples regarding whether a family history revealing a number of blood relatives who have been alcoholic or depressed, or have suicided, or in other ways have been afflicted with a mental illness may manifest itself if a child is conceived, and to what degree social and psychological (nonbiogenetic) factors may reduce manifestation of this gene.

The information derived must be reviewed and critically presented in much the same way one would present data on transmission of insulin-dependent diabetes or cystic fibrosis. If a method were ever available that would allow detection of a child at high risk for autism or schizophrenia prenatally in much the same way Down's syndrome can be identified through amniocentesis, a mother, if concerned, should be made aware of the availability of such technology and offered the choice to know of the potential risk and of the factors that may diminish or eliminate that risk. Ultimately, the goal of psychiatric epidemiology is the provision of understanding of variables impacting the manifestation of illness so that effective programs of prevention and treatment can be developed.

In some instances there may be patterns of response transmitted that may in some individuals prove maladaptive while in others prove highly adaptive. For instance, a number of the contributors, in particular Michael Lyons and his coauthors, discuss the multiple factors contributing to the manifestation of schizophrenia and how there may in fact be variants of presentation of this illness. James Joyce, the great Irish writer, had a daughter felt to be schizophrenic who was treated by the eminent Swiss psychiatrist Carl Jung. Jung, it is reported, was asked by Joyce what was the difference between his writing and that of his daughter. After all, doesn't *Ulysses, Dubliners,* or *Finnegans Wake* in part recall the loose associations used in the diagnosis of "thought disorder"? Jung reputedly replied to Joyce's question: "It is the difference between jumping in and drowning and jumping in and coming up again." We would not like to lose the brilliant contributions of people like James Joyce more than we would wish the pain of schizophrenia (associated with a high suicide rate) on anyone.

In the chapter by Brennan and her associates (Chapter 12) the possible relationship between transmission of criminality and responsiveness and recovery of the autonomic nervous system is discussed. This fact, if consistent, may mean that some people who commit crimes do not "feel guilty" or learn from their mistakes in the same way the rest of us may. This tendency, however, may be adaptive in performing creative tasks that other people cannot because they are overwhelmed with stress or are afraid to take risks because of "what people will say."

Finally that which predicts morbid outcome may ironically also predict extreme success. Cite, for illustration, the relationship between creativity, suicide, and affective illness. In this case, it appears the common factor is decreased levels of central nervous system serotonin and enhanced proclivity for impulsive action. Dr. Marie Asberg and her colleagues in Sweden and investigators elsewhere have found that violent suicides (i.e., those who die by jumping, hanging, shooting, cutting their throats, or other usually lethal means as opposed to drug overdose) have reduced serotonin in their brains. Comparably, when comparing the levels of 5-hydroxy-indoleacetic acid in the cerebrospinal fluid of those who make violent suicide attempts to those who overdose or use less lethal means and to normal controls, there is a decreased concentration of this metabolite of serotonin. She concludes that decreased serotonin in the central nervous system is not a finding in those who attempt suicide as much as it is in specifically those who make violent attempts. Interestingly, when examining the cerebrospinal fluid of individuals who have committed impulsive violent homicides we also find a decrease in the breakdown product of serotonin. It is this group that also shows the changes in autonomic nervous system activity that Brennan et al. discuss in their chapter on criminality. The creativity also seen in excess in some groups of affectively ill patients (Slaby, forthcoming) may in part relate to what may be a genetic tendency to impulsivity and not simply to affective illness. Just as this group may impulsively feel that the hopelessness experienced when severely depressed will never go away and they may violently end their lives, so too when they have an excellent new idea for a book, painting, business venture, or political change they may impulsively go with it. They may not obsess as much as those who respond differently autonomically.

It is the complexity of the relationship among genetically determined factors, social press, and developmental variables that makes this volume special. Tsuang, Kendler, Lyons and their co-contributors present state-of-the-art material on genetic epidemiology in a critical and responsible manner and integrate it with data from other areas of research. One feels a dynamism of interaction in reading this and senses the pace at which new discoveries are being brought to bear on our assumptions. This book really has both scientific and social implications. We may find to none of our surprise that our vulnerabilities biogenetically may also be tied to our strengths. Our challenge will be how we may realize our natural strengths and talents to their fullest degree without perishing from the attendant pain. This volume provides some clues on how we may do this.

A.E.S.

See Slaby, A. E. (forthcoming) Psychopharmacotherapy of suicide and self-destructive behavior. In *Dangerous Intersections: Assessment and Management of Suicidality in Clinical Practice*, ed. Bruce Bongar. New York: Oxford University Press.

Genetic Issues in
Psychosocial Epidemiology

Part 1

Overview and Methods

Chapter 1

Introduction

KENNETH S. KENDLER
MICHAEL J. LYONS
MING T. TSUANG

The conceptual framework of this book emerges from combining the perspectives of medical genetics on the one hand and epidemiology on the other. Traditionally, medical genetics has focused on disorders that were the result of a single abnormal gene, often called an SML (for *single major locus*). Such disorders, termed *Mendelian* after the founder of modern genetics, Gregor Mendel, have several important properties. First, they are rare, that is, they usually occur in less than one in a thousand individuals and often in less than one in ten or fifty thousand. Second, in the etiology of these disorders, environmental risk factors are inconsequential: individuals carrying the necessary abnormal gene become affected regardless of their environmental experience. Third, there is rarely a problem in distinguishing affected from unaffected individuals; and fourth, the disorder cannot usually be mimicked by nongenetic factors. Thus, the *genotypes*, or genetic constitutions, of the individuals (do they or do they not have the disease gene?) can be accurately inferred from their *phenotypes*, or clinical conditions (are they ill or well?). Finally, there is with many of these Mendelian disorders strong a priori evidence that most or all cases of the illness derive from abnormalities in the same gene (i.e., there is genetic homogeneity). Disorders of this kind demonstrate one of a small number of clear patterns of genetic transmission. These "Mendelian" patterns such as autosomal dominant (never skips a generation, and approximately 50% of the siblings and offspring of affected individuals are affected) and autosomal recessive (there is approximately 25% risk to siblings and usually a very low risk to parents and offspring) are clearly demonstrable in pedigrees and can easily be recognized without the assistance of complex computer algorithms.

Historically, epidemiology and medical genetics have viewed

medical illness from quite different perspectives. While medical genetics has focused solely on internal causes of illness, epidemiology has concentrated its attention on external, or environmental, causes. Whereas medical genetics has always studied individuals in the context of their family units, epidemiology has generally focused on the distribution of illness in individuals without regard to family structure. Traditionally, medical genetics has ignored the etiologic importance of environmental factors in human illness; epidemiology has ignored the etiologic importance of genetic factors.

From these disparate and sometimes conflicting traditions, the new discipline of genetic epidemiology has emerged (Morton, 1982). This discipline, which combines elements from both traditional medical genetics and epidemiology, focuses on disorders with the following characteristics: (i) they are common (with lifetime risks usually in excess of 1%); (ii) they tend to run or aggregate in families, (iii) they have no clear Mendelian pattern of transmission, (iv) the boundary between affected and unaffected individuals is often unclear, (v) the impact of genetic factors is modulated by environmental risk factors, (vi) the disorder can be produced by environmental factors in individuals with low genetic risk, and (vii) for many of the disorders, it is probable that it may be produced by a number of different genes (genetic heterogeneity).

What disorders have these characteristics? The list is a long one and would include all of the major medical conditions of Western societies, such as heart disease, hypertension, diabetes, peptic ulcer disease, autoimmune and allergic disorders, and some forms of cancer—as well as all of the major psychiatric disorders, including schizophrenia, bipolar and unipolar affective disorder, alcoholism, panic disorder, generalized anxiety disorder, phobias, obsessive-compulsive illness, and antisocial personality.

The task of genetic epidemiology can best be illustrated by setting forth the pertinent issues. First, what are the nature and magnitude of the familial aggregation of a disorder? This apparently simple question can be quite difficult to answer conclusively. Although we cannot here examine in detail the methodologic issues in what are termed "family studies," it is important to note that the value of the results of such studies and the strength of the conclusions to be drawn from them depend upon the rigor of the experimental methods employed. Firm conclusions require an appropriately ascertained group of affected probands (individuals identified, or "ascertained," through treatment facilities or psychiatric registries, and whose families will then comprise the subjects of the study), one or more carefully matched control groups, adequate and unbiased assessment methods in relatives, and blind diagnostic evaluations (in which the diagnostician is unaware whether the relative is related to an ill or control proband). Many family studies in the psychiatric literature suffer from inadequate sample sizes, lack of a control group, affected probands obtained as "samples of convenience" rather than as representatives of the universe of affected individuals, and insensitive assessment methods that rely on indirect reporting by one individual on other members of the family (the so-called family history method) instead of direct assessment of each individual relative (the so-called family study method).

The second major task for genetic epidemiology, and one that is particularly relevant for psychiatry, is to understand the syndromes that are being transmitted in families. In other words, if one begins with a set of probands with disorder A, what other disorders will be found in their close relatives in rates higher than those in the general population. Will it only be disorder A or will it also be disorders B and C? An example might clarify the importance of answering this question. Assume we performed a family study of hemorrhagic stroke, unaware that hypertension was the major risk factor for this disorder. Beginning with a series of probands with hemorrhagic stroke, we found in their relatives an excess of hemorrhagic stroke *and* an excess of several other diseases, including congestive heart failure, myocardial infarction, hypertensive renal disease, and so on. It might then occur to us that what is really transmitted in such families is hypertension.

A similar approach has been taken for most psychiatric disorders. As reviewed by Lyons et al. in this volume, there is considerable evidence that a nonpsychotic schizophrenia-like personality disorder (termed "schizotypal personality disorder") aggregates in relatives of schizophrenic probands. As discussed by Winokur, relatives of patients with bipolar affective illness (manic-depressive illness) are at increased risk for episodes of depression (unipolar illness). Uncovering these patterns of illnesses in families (coaggregation) will help us to understand the etiologic relationship among psychiatric disorders. In addition, more sophisticated genetic studies of psychiatric illness, such as linkage and segregation analysis, often require that relatives be classified as affected or unaffected. Studies of coaggregation are crucial in determining affection status (affected versus unaffected).

The third major issue in genetic epidemiology is the need to determine why a disorder aggregates in families. Does this occur because relatives share environmental risk factors or because they share genes? For example, someone interested in developing methods for preventing a disease would employ different strategies depending upon whether the major etiologic factors involved were genetic or environmental. Many different kinds of environmental factors might be responsible for the familial transmission of a disease. Certain disorders might be transmitted directly from parent to child in much the same manner as language, religious affiliation, or dietary preferences. For siblings, environmental factors that might cause similarity in disease risk include parental rearing styles, religious and cultural factors, social class variables such as schools and neighborhoods, and exposure during childhood to toxins or infectious agents. Genetic factors might include genes of major effect (SMLs) or, alternately, many genes each exerting a small effect on disease vulnerability (polygenes).

Few areas in psychiatric genetics have incited so much controversy as the question of why disorders aggregate in families. For schizophrenia, in particular, investigators from divergent theoretical backgrounds have interpreted the same data in diametrically opposed ways. Those who were convinced that schizophrenia was a psychogenic disorder produced by disturbed family relations *assumed* that the familial aggregation of schizophrenia was the result of familial-psychological factors. Those who believed schizophrenia to be a "biological" illness *assumed* that the familial aggregation of schizophrenia resulted from genetic transmission. Fortunately, this theoretical controversy is giving way to a more scientific approach to the problem: the dispassionate and critical use of all methods available to disentangle genetic and nongenetic sources of familial resemblance.

In most instances, individuals share many environmental factors with their close biological relatives, with whom they also share many genes. Therefore, in family studies, genetic and environmental sources of familial resemblance are confounded. In attempting to unconfound the effects of genes versus shared environment, psychiatric genetics has traditionally made use of two natural experiments: twin studies and adoption studies. The methodology of twin and adoption studies is discussed in separate chapters by Rose and Cadoret, respectively. In addition, the chapter by Segal et al. provides an example of the rare study that combines both methods in studying twins reared apart. It is important to note here the differing strengths and limitations of twin and adoption studies. As discussed by Rose, twins tend to have a less ideal gestational and birth experience than does the nontwin population. Genetic models for the analysis of twin studies assume that the environmental experiences of identical twins are no more similar than the environmental experiences of fraternal twins. This appears to be true for many traits, but not for all. Therefore, it is important to determine under what circumstances the "equal environment assumption" of twin studies is valid.

Theoretically, adoption studies provide an even cleaner separation between the effects of nature and nurture than do twin studies. However, while twins, aside from excess rates of birth complications and slightly slower language development, are both common and representative of the general population, adoptees are becoming increasingly rare and are often atypical in important ways. A major task of adoption agencies is to screen prospective adoptive parents and eliminate those with major-psychopathology. Often, the biologic parents of adoptees are atypical in the opposite way, having excess rates of psychopathology—especially alcoholism. Finally, for some years, adoption agencies practiced "selective placement," an attempt to match characteristics in the biologic and adoptive parents. This could, in theory, entirely confound the results of adoption studies.

Because of their complementary strengths and weaknesses, when evaluating the important insights into the causes of familial aggregation of psychiatric disorders provided by twin and adoption studies, one is on firmer ground when results for any given disorder can be compared across both methods.

It should be mentioned that the increased power of linkage analysis, which examines in families the tendency for a disease and a marker that tags one part of the human genome to be inherited together, has provided a new tool to clarify the mechanism of the familial transmission of psychiatric disorders. It is a subject of considerable debate whether linkage analysis in a familial disorder (one that has been shown to aggregate in families) should await the production of strong evidence from twin and/or adoption studies demonstrating a strong genetic effect. It is at least clear that the successful demonstration of linkage for a disorder would prove the etiologic importance of genetic factors even more conclusively than would twin or adoption studies.

A fourth major problem in genetic epidemiology is to try to understand the interrelationship between genetic and environmental risk factors for disease. This is a particularly important area because it can provide insights into ways in which disorders can be prevented. Two conceptual models are gene-environment interaction and gene-environment correlation (Kendler & Eaves, 1986). Gene-environment interaction means that individuals with varying genotypes will differ in their sensitivity to environmental risk factors. This principle may be more meaningfully termed "genetic control of sensitivity to the environment." For example, the effect of sodium intake on blood pressure varies widely among individuals, and this variability is in part under genetic control (Kawasaki et al., 1978). Because of their genotypes, some individuals are more sensitive than others to the hypertensive effects of a high sodium diet. A similar mechanism can be postulated for many psychiatric disorders. For example, could genes for depression function in part to increase sensitivity to the depressogenic effects of early parental loss?

Another way to conceptualize gene-environment interaction is that environmental experiences may "switch on" genes that otherwise lay dormant with respect to a given phenotype. For example, certain chemicals, called "precarcinogens," do not themselves cause cancer but they can be metabolized into carcinogenic agents. Because of genetic differences, some individuals convert these chemicals into their carcinogenic products with great efficiency and others do not. Thus, without exposure to these precarcinogens, these genes are dormant with respect to the risk for cancer. However, given chemical agents, they then become "switched on" and can potently affect the risk for the development of cancer. Is it possible that analogous situations exist in psychiatric genetics? Might certain unusual experiences, such as combat exposure, "switch on" genes for behavioral repertoires that are otherwise unexpressed in civilian life, thus greatly effecting the risk for the development of syndromes such as post-traumatic stress disorder?

Gene-environment correlation, which may more meaningfully be called "genetic control of exposure to the environment," occurs when the probability of exposure to an environmental risk factor is genotype dependent. For example, genetic factors have been implicated in cigarette smoking (Eysenck & Eaves, 1980). Given that smoking is such a potent risk factor for lung cancer, genes for smoking in fact function as genes for lung cancer. These "smoking genes," however, do not directly increase the risk for cancer. Rather, they work by making it more probable that individuals will expose themselves to a high risk environment. We and others (Kendler et al., in review; Plomin et al., 1991) have shown that life events are influenced by genetic factors. Thus, genes for psychiatric illness may work in part not by directly increasing risk for illness but rather by increasing the probability that an individual will experience environmental risk factors, such as traumatic life events.

The fifth and final major task in genetic epidemiology is to determine the mode of transmission of the genetic vulnerability to the disorder in question. Although there are many different modes of genetic transmission, most effort in genetic epidemiology has been concentrated on distinguishing between genes of major effect (SMLs) and genes of minor effect (polygenes). The difference between these two modes of transmission may best be illustrated by examples. SMLs cause classic genetic diseases such as cystic fibrosis, Huntington's chorea, muscular dystrophy, and phenylketonuria. They are also responsible for many human discrete traits such as blood groups (e.g., ABO, Rhesus) and eye color. Other genetically influenced traits, including physical traits like height, weight, and skin color (and probably also such psychological traits as personality and intelligence), appear to result from the action of a number of genes, each of which makes only a relatively modest contribution. It should be noted that the distinction between SMLs and polygenes is itself quantitative rather than qualitative. In fact, as few as three to five genes can produce a pattern of transmission that is indistinguishable in realistic sample sizes from that found with classic polygenic systems (Kendler & Kidd, 1986).

If SMLs influence psychiatric disorders, they do so in ways that are significantly different from those for classic Mendelian disorders. Most importantly, the probability of having the disease if you have the disease gene or genes, termed the *penetrance*, is not unity. That is, one can possess the disease gene but not manifest the illness. This difference is most clearly demonstrated by examining identical (monozygotic or "MZ") twins, who, because they develop from the same zygote (fertilized egg), have identical genotypes. With classic Mendelian diseases, if one member of an MZ twin pair has the illness, the co-twin will always be affected. However, for psychiatric disorders, it is common to find MZ twin pairs where one has the disorder and the other does not. Therefore, if there are major genes for psychiatric illness, they are almost certainly *nonpenetrant* in some individuals.

Our power to discriminate among different modes of genetic transmission has increased dramatically in the last decade as a result of parallel advances in molecular biology and statistical genetics. As outlined in the chapters by Risch and O'Donovan and McGuffin, the advances in molecular genetics have provided us, for the first time, with the realistic possibility of identifying major genes for psychiatric illness. However, this effort is likely to prove many times more difficult than finding linkage for Mendelian diseases, an area which has recently boasted many spectacular successes. In fact, no one has yet uncovered conclusive evidence of the existence of major genes for psychiatric disorders. Also, it is still unclear whether the problems posed by genetic heterogeneity will prove to be minor (there are two or three genetic forms of illness, each of which comprises a substantial proportion of familial cases) or major (there are tens or hundreds of different genes with no single gene alone causing even a substantial minority of cases). In the latter instance, the problems for linkage studies may be so great as to prove insurmountable.

The results of linkage studies of psychiatric disorder to date have proved frustrating. For example, the initial highly publicized strong evidence for linkage of manic-depressive illness in the Amish to markers on chromosome 11 (Egeland et al., 1987) could not be replicated in the same pedigree in which it was initially found (Kelsoe et al., 1989). Many attempts to replicate the linkage of schizophrenia to chromosome 5 markers (Sherrington et al., 1988) have to date proved unsuccessful.

Looking Toward the Future

Although speculating about the future is hazardous, it is also sometimes difficult to resist. One of the most important issues to face psychiatric genetics in the next decade is whether the substantial effort and considerable resources now being expended on mapping single genes for mental illness will bear fruit. If it produces replicable results, a whole new vista will open up for psychiatric genetics and, indeed, for the biology of mental illness. For example, through the application of what has been termed "reverse genetics," it may be possible, through linkage analysis, to directly identify the abnormal gene and thence to unravel the pathophysiology of the disorder. Should this effort prove unsuccessful, however, we must be prepared for a potential backlash of sentiment against the field. But failure to uncover the mechanism of genetic transmission of psychiatric disorders with current methods of linkage analysis, which are capable of detecting only genes of major effect in the absence of high levels of genetic heterogeneity, may mean only that the underlying liability to these disorders is polygenic in origin or that the disorders are very genetically heterogeneous; more powerful molecular and statistical methods may be needed to detect genes under these conditions.

The field is currently in some danger of being blinded by the bright lights of molecular genetics. Although they are unlikely to lead to the kind of dramatic breakthrough that would occur with successful linkage, many other efforts in psychiatric genetics will continue to increase incrementally our knowledge of psychiatric disorders. Statistical methods for understanding gene-environment interactions and correlations are improving and field studies that attempt to assess, in a single sample, both genetic and environmental risk factors are becoming more common. Multivariate genetic techniques will allow us to understand with much greater precision the relationship between the familial vulnerabilities to various psychiatric disorders. That is, while epidemiology can only observe comorbidity (the tendency of disorders to occur together in the same individual), psychiatric genetics can take a critical step beyond that observation and determine whether disorders tend to cooccur because they are influenced by the same environmental risk factors, the same genes, or both.

Another important area of research is the attempt to identify psychological, neurophysiological, or neuropsychological traits that are potential indices of the familial or genetic vulnerability to psychiatric illness. As illustrated in the chapter by Matthysse, smooth pursuit eye movements are one such potential index. It is possible that this line of research will yield important insights into the nature of the genetic liability to psychiatric disorders.

Psychiatric genetics will also continue to interact synergistically with psychiatric nosology: nosology is a critical tool for psychiatric genetics, and psychiatric genetics remains one of the most important empirical methods available to refine our nosologic categories.

Finally, it is important that psychiatric genetics not lose sight of its important epidemiologic roots. Although genetic factors may be of nearly overwhelming importance in some psychiatric disorders (e.g., schizophrenia and bipolar illness), the best evidence suggests that they may be of rather less importance for such disorders as major depression, alcoholism, and many personality disorders. Complete genetic analyses of these conditions may require the incorporation of epidemiologic observations such as cohort effects (see chapter by Rice et al.). Understanding the etiology of psychiatric disorders will require a knowledge of the environmental risk factors, the genetic risk factors and, perhaps most importantly, the interaction between them.

References

Egeland, J. A., Gerhard, D. S., Pauls, D. L., Sussex, J. N., Kidd, K. K., Allen, C. R., et al. (1987) Bipolar affective disorders linked to DNA markers on chromosome 11. *Nature* 325:783–787.

Eysenck, H. J. & Eaves, L. J. (1980) *The Causes and Effects of Smoking*. Beverly Hills, CA: Sage Publications.

Kawasaki, T., Delea, C. S., Bartter, F. C. & Smith, H. (1978) The effect of high-sodium and low-sodium intakes on blood pressure and other related variables in human subjects with idiopathic hypertension. *American Journal of Medicine* 64:193–198.

Kelsoe, J. R., Ginns, E. I., Egeland, J. A., Gerhard, D. S., Goldstein, A. M., Bale, S. J., et al. (1989) Re-evaluation of the linkage relationship between chromosome 11p loci and the gene for bipolar affective disorder in the Old Order Amish. *Nature* 342:238–243.

Kendler, K. S. & Eaves, L. J. (1986) Models for the joint effect of genotype and environment on liability to psychiatric illness. *American Journal of Psychiatry* 143:279–289.

Kendler, K. S. & Kidd, K. K. (1986) Recurrence risks in an oligogenic threshold model: The effect of alterations in allele frequency. *Annals of Human Genetics* 50:83–91.

Kendler, K. S., Neale, M. C., Kessler, R. C., Heath, A. C. & Eaves, L. J. (in review) A twin study of recent life events and difficulties: The genetics of the "environment."

Morton, N. E. (1982) *Outline of Genetic Epidemiology.* New York: Karger.

Plomin, R., Lichtenstein, P., Pedersen, N., McClearn, G. E. & Nesselroade, J. R. (1991) Genetic influence on life events during the last half of the life span. *Psychology and Aging* 5:25–30.

Sherrington, R., Brynjolfsson, J., Petursson, H., Potter, M., Dudleston, K., Barraclough, B., et al. (1988) Localization of a susceptibility locus for schizophrenia on chromosome 5. *Nature* 336:164–167.

Chapter 2

Twin Studies and Psychosocial Epidemiology

RICHARD J. ROSE

It is a truism that both genes and experience contribute to individual differences in behavior, life-style, and disease risk. Few dimensions of health behavior are insulated from genetic expression, but the magnitude and timing of genetic influences differ across behaviors and life-styles, and are modulated by the complex interactions of age, gender, and familial history of disease. Conversely, perhaps no dimension of health behavior is independent of variation in experience, but identification of environmental effects is invariably confounded with genetic differences between individuals, so an incisive appraisal of environmental factors in disease risk requires research designs that eliminate genetic confounds. Analyses of differences within and between human twins, analyses first suggested more than a century ago by Sir Fráncis Galton, provide fundamental data in resolving the causes of familial and individual variation in behavioral development and behavior disorder.

In 1965, a collaborative research group, meeting under the auspices of the World Health Organization, reviewed the potential of twin methods in epidemiological research (WHO, 1966). In the quarter-century since that review, assumptions underlying twin methods have been evaluated, variations of the classical twin model have been developed, extensive new twin registries have been organized, and applications of biometrical techniques to twin and family data have been made. This chapter will survey assumptions, methods, and representative results of twin studies as research tools in psychosocial epidemiology.

Restrictive Assumptions on the Use of Twins

Underlying the use of twins in epidemiological research is a restrictive assumption that twins are representative of the general

12

population of nontwins to whom inferences are to be made. The degree to which valid inferences can be drawn from twin data depends, obviously, on comparability of twins to nontwins with regard to the developmental history of the disease or trait under study. Twins systematically differ from singletons in several potentially significant ways, and it must be demonstrated that these systematic differences are not causally or etiologically relevant. Circumstances of twin pregnancy and delivery, and prenatal and postnatal environmental influences unique to twins, must be evaluated with appropriate tests.

First, factors associated with the twinning process may also be associated with the trait or disease under study. Dizygotic (DZ) twinning, a consequence of polyovulation, is related to maternal age and parity, as are many congenital disorders of interest to epidemiologists. The placentation of most monozygotic (MZ) twins differs from that of DZ twins or singletons, and placental differences may relate to variables of epidemiological importance. Birth weight and gestational age of twins systematically differ from those of singletons, and early in development these factors predict motor and mental growth. Tests of these potential biases must be made by evaluating the prevalence of the disease (or equivalence of the trait) in twins relative to singletons and by testing the generalizability of inferences drawn from twin data to results obtained from other data sets, including nuclear families of nontwins.

Second, experiences unique to being a twin may influence the early development of some behavioral traits or psychopathologies. Being a twin may significantly bias postbirth social experience. Thus, it was suggested that experience unique to MZ twins impairs development of their personal identity and creates an elevated risk for schizophrenia (Jackson, 1960). Again, the obvious test is a comparison of the prevalence of a particular trait or disorder among twins in general, and monozygotic twins in particular, with that of singletons. For schizophrenia, these risks are equivalent (Kendler & Robinette, 1983; Rosenthal, 1960). Twins do differ from singletons in some aspects of early behavioral development, including language fluency, but such differences dissipate with age, and by puberty few consistent differences remain. Results from a United States study of 850 adolescent twin pairs (Loehlin & Nichols, 1976) indicate that on standardized personality, interest, and aptitude tests, twins obtain scores that are indistinguishable from those of nontwins; on scales of the California

Personality Inventory, means and variances for the 1,700 twin subjects were virtually identical to norms derived from the responses of nontwins. In the face of such evidence, it appears highly unlikely that experiences peculiar to twins have systematic effects on personality development.

Third, a pervasive problem in twin research in the United States and the United Kingdom is introduced by self-selection, because nearly all twin studies outside Nordic countries have recruited volunteer twins. Analyses (Lykken et al., 1978; Lykken et al., 1987) suggest systematic bias in volunteer studies, because two-thirds of a typical sample is female and two-thirds of a typical sample is monozygotic. This "rule of two-thirds" suggests that those dizygotic male twins who do volunteer may be particularly unrepresentative of the population. Volunteer bias may alter means of traits related to the tendency to volunteer and, by decreasing variance in DZ twins, may bias estimates of genetic parameters (Martin & Wilson, 1982). Fortunately, major twin research efforts in Nordic countries have utilized population registries that eliminate problems of biased ascertainment (Kaprio et al., in press) and efforts are under way to develop such registries in the United States, from statewide birth records in Minnesota (Lykken et al., 1990) and Missouri and motor vehicle records in Virginia.

The Classic Twin Method

Assumptions

In addition to assumptions necessary for valid generalizations from twins to nontwins, a second set of assumptions is necessary for valid comparisons of the two twin types. Those assumptions have been challenged frequently, and in response systematic tests for their evaluation have been developed. Results are reassuring.

The first assumption underlying the classical twin method is that of equality of means and variances. It is critical to determine whether the level of variability of the trait under study is associated with twin type, because some recent evidence suggests that it is not uncommon to find such associations. For quantitative traits, appropriate t' and F' tests for evaluating equality of means and variances can be used (Christian, 1979); for qualitative disorders, assumptions of equal prevalence can be evaluated by testing the difference in relative frequencies of affected individuals in representative samples of MZ and DZ twins.

The second assumption necessary to the classical twin method is equality of environmental covariance in MZ and DZ twin pairs. No aspect of the twin method has been challenged more frequently than this assumption of "equal environments," and in recent years no assumption has been tested more rigorously. MZ twins do have more similar experiences than do DZs. They are more likely to dress alike, more likely to sleep in the same room during childhood, more likely to play together, more likely to share a similar circle of friends, and so forth. The question of interest is whether these factors systematically influence personality development. To answer that question, Loehlin and Nichols (1976) correlated within-pair differences in experience with differences in quantitative dimensions of personality. Thus, for example, some identical twins dress alike, but others do not. Within a sample of identical twins we can test whether the tendency to dress alike is associated with similarity in personality. Extensive tests by Loehlin and Nichols suggest a negative answer for that aspect of early experience and for others tested.

There is, however, another approach to the issue. Adult twins vary in the amount of social contact they maintain with one another, and MZ twins have more frequent interaction than do age-matched DZ pairs. For questionnaire measures of neuroticism (Rose et al., 1988) and for self-reports of social drinking patterns (Kaprio, Koskenvuo et al., 1987), twins in more frequent contact were found to be more alike. These effects, although statistically significant, were modest in magnitude, and genetic influences remained significant when tested, with stepwise regressions, after effects of social contact were first removed (Kaprio, Koskenvuo et al., 1987; Rose et al., 1988).

A second test of equality of environmental covariance correlates similarity of appearance with similarity of personality. Identical twins obviously look more alike than do fraternal twins. Perhaps they elicit more similar histories of reinforcement from the social community and create more similar opportunities for their behavioral development. Is this a factor in their greater personality resemblance? Plomin et al. (1976) evaluated similarity of appearance in MZ and DZ twins and, as expected, found large differences associated with zygosity. But within MZ twins, no correlation between similarity of appearance and similarity in personality was found, and that finding has been replicated in very young twins as well (Matheny et al., 1976).

A third approach to the issue, reported by Lytton (1977), concerns interactions of parents with their 2½ year old twin sons. Parental interaction with MZ co-twins was significantly more similar than with DZs, but this was a consequence, rather than cause, of the differences in twins' behavior. Distinguishing parent-initiated from child-initiated actions, Lytton concluded that MZ twins are treated more alike by their parents, because they provide more similar cues for parental response.

A final test of the equal environments assumption, introduced by Scarr (1968) and replicated by Matheny (1979), studies twin pairs misclassified for zygosity. Many twin children are misidentified for zygosity by themselves and their parents, usually on the basis of erroneous placental information at the time of delivery. Behavioral differences in such misclassified twin pairs relate to their true zygosity differences, not to their parents' beliefs regarding zygosity.

In short, equality of environmental covariance remains a tenable assumption for epidemiological research. Other than differences in frequency of adult social interaction or duration of cohabitation, there is no consistent evidence of important differences in experience among MZ and DZ twins. In their large sample of adolescent twins, Loehlin and Nichols (1976) found that only 5 of 1,600 discrete variables reliably distinguished the two twin types. And only a single item consistently differed in MZ/DZ twins of both sexes. That item? Pretending to be one's twin!

Representative Results

When its necessary assumptions are met, the classic comparison of MZ and DZ twins offers an incisive tool for epidemiological research. For illustrative purposes in the limited space here available, twin research on the schizophrenias will be cited. Three decades ago, familial aggregation of schizophenia was attributed to pathogenic elements in familial social environments. No longer. The decade of the 1960s witnessed a remarkable volume of research using the classic twin comparison, illustrated by twin studies of schizophrenia in Japan (Inouye, 1963), the United Kingdom (Gottesman & Shields, 1966), Norway (Kringlen, 1967), Finland (Tienari, 1968), Denmark (Fischer et al., 1969), and the United States (Pollin et al., 1969). A general conclusion required by this body of twin research is that genetic factors play a role in the predisposition to schizophrenia, and that fact has profoundly altered current research strategies. Twin and adoption data collected since the early 1960s have fostered high-risk research and anterospective studies in which subjects are identified at risk via their genetic relationship to an affected proband.

Were this the only result classic twin methods offered psychosocial epidemiology—a convincing demonstration that genes play a role in the causal chain—it would be a significant contribution. But demonstrations of heritability are merely a first step. Twin study data are critical in efforts to determine the developmental pathways and nature of gene action, to assess the nature of common (familial) and specific (individual) environmental effects, and to document the presence of gene-environment interactions and correlations. The end results, reached through applications of biometric genetic analyses, can guide the direction of contemporary theory and research.

Biometric Genetic Analyses

An influential paper by Jinks and Fulker (1970) first applied the biometric techniques of animal and plant genetics to twin data. The starting point for the analysis is a partitioning of total behavioral variation into within- and between-family components. Within-family environmental variance, created by specific experience (SE), is idiosyncratic to each family member; illustrated by birth order, SE makes siblings differ. It is distinguished from between-family environmental variance, a product of a family's common experience (CE); illustrated by social class, CE makes siblings alike. Because MZ co-twins are genetically identical, all differences between them must be environmental, and intrapair differences of MZs reared together (MZT) provide an estimate of SE. An estimate of CE can be obtained from the greater similarity of MZs reared together relative to those reared apart (MZT − MZA), while the latter, MZA, provides a direct estimate of genetic variance. So, with the usual assumptions and with correlations (in this case, for negative emotionality scale scores from

Tellegen et al., 1988) of MZT = .61 and MZA = .54, we estimate h^2 = .54, SE = (1 − MZT) = .39 and CE = (MZT − MZA) = .07.

Extensive analysis of a questionnaire measure of neuroticism was reported by Eaves (1978) based on data from nearly 2,500 individuals including 543 twin pairs and 340 fostered children. Parameter estimates of genotype-environment models were obtained by maximizing the likelihood of observed pedigree data. The estimates permit inferences similar to those made from likelihood tests based only on results from twins. Surprisingly, yet consistently, there was little evidence that common family experience contributes to personality variation, a finding confirmed in model-fitting the largest and most recent twin study data on the personality dimensions of extraversion and neuroticism (Loehlin, 1989). Biometric analyses of schizophrenia yield similar results: studies based on carefully defined twin registries reveal that concordance for schizophrenia among MZ twins is 3 to 5 times that of DZs. Data from the United States panel of veteran twins (Pollin et al., 1969) illustrate that fact: the panel consists of 31,818 veteran twins for whom unique calculations of population prevalence rates can be made and compared to risks for co-twins of schizophrenic probands. Early results, based on file records, yielded an MZ/DZ concordance ratio of 3.3; prevalence of schizophrenia in the population of veteran twins approximates 1% in both MZ and DZ twins, demonstrating again that being an MZ twin is not associated with the risk of being schizophrenic; but schizophrenia was 25 times more frequent for MZ co-twins of schizophrenics than in the population of MZ twin veterans, whereas prevalence in DZ co-twins of schizophrenics was 6.5 times its population prevalence (Gottesman & Shields, 1972). Updated information on this registry, based on a 16-year follow-up (Kendler & Robinette, 1983), again found recorded diagnoses of schizophrenia equally common in MZ and DZ twins, but the MZ/DZ concordance ratio was 4.75; neither biases in ascertainment, diagnosis, nor zygosity determination could plausibly explain this concordance ratio, permitting the conclusion that genetic factors are as important in schizophrenia as in diabetes or hypertension.

But if a genetic diathesis is necessary for the development of true schizophrenia, it is not sufficient: a significant causal role must be assigned to environmental influences, because the pooled concordance of MZT is less than 50%. Yet, the causally relevant environment is one idiosyncratic to family members rather than that shared by them: concordance of MZA equals that of MZT, and fostering offspring of a schizophrenic parent fails to reduce the risk of schizophrenic outcome. For schizophrenia, as for many dimensions of normal personality, a simple model of additive genetic variance, combined with effects of specific experience, adequately fits data of twin-family aggregation. That no significant role can be attributed to a common family environment must constitute one of the most important facts we possess about the meaning and nature of schizophrenia. New evidence that offspring of schizophrenic parents are at risk because of transmitted genes, rather than deviant parenting, has been found in studies of the offspring of twin parents discordant for schizophrenia. These new data will be described after the families-of-twins research design, on which the data are based, is introduced.

Extended Twin–Family Research Designs

Variations on the conventional twin comparison include extended twin-family research designs in which twins' families, rather than twin pairs, comprise the sampling unit. Two complementary research strategies are now common. The first samples adolescent or young adult twins and their parents; covariance matrices obtained from these balanced pedigrees permit robust estimates of genetic and environmental parameters. An illustration, using self-report data on common fears and phobias, combined twin-parent pedigrees with those from nontwin siblings and their parents (Phillips et al., 1987) to obtain maximum-likelihood estimates of genetic and cultural effects; results were consistent with earlier evidence, from conventional twin comparisons, that genetic predispositions, and consequent prepared learning, underlie common fears (Rose & Ditto, 1983; Rose, Miller et al., 1981). Twin-parent data of this kind can be obtained efficiently, and simulation studies reveal the power of the data structure in the resolution of heritable and cultural variance (Heath et al., 1985).

A second strategy samples kinships of twin parents. Adult MZ co-twins with their spouses and children provide unique data for epidemiological research. Because either their fathers or mothers have identical nuclear genes, children of MZ co-twins are genetic half-siblings, but they are reared as cousins in different house-holds. Accordingly, these MZ kinships permit an incisive separation of shared environments from shared genes. One can compare the resemblance of full siblings reared in a common home to that of genetic half-sibs who are reared apart; children in each kinship are genetically related to their twin uncle or aunt, the MZ co-twin of their twin parent, as closely as they are to their own father or mother. Each child shares a household environment with his father or mother, but not with his twin uncle or aunt. Finally, the child's resemblance to his nontwin uncle or aunt, the spouse of his father or mother's MZ twin, reflects neither shared genes nor common home environment, and in the absence of assortative mating, no resemblance is expected. Thus, the multiple genetic and social relationships within kinships of MZ twin parents permit a systematic assessment of the influence of common home environment in paired individuals who share all (the twins), one-half (e.g., the full sibs), one-quarter (the half-sibs), and no (e.g., the spouses) genes identical-by-descent. The design was illustrated (Rose, Harris et al., 1979) in an analysis of data from the Block Design subtest of the Wechsler Intelligence Scales. Results paralleled those found for fingerprint ridge count, a metric trait formed in early prenatal life and immune to postbirth experience. Familial resemblance for both traits was shown to be a direct function of shared genes, and neither cultural effects nor assortative mating was evident. The simple model of additive genetic variance together with effects of specific experience adequately fits observed data. Similar results were found (Rose, Miller, Grim & Christian, 1979) for systolic blood pressure, and a biometrical analysis of that trait (Rose, Fulker et al., 1980) was made from a nested analysis of variance of the twin and half-sib data. Inferences drawn from that analysis are consistent with data from a recent report that included kinships of both MZ and DZ twin parents (Tischler et al., 1987).

The absence of common environmental influences on blood pressure is of interest, given conventional wisdom that salt ingestion, exercise patterns, and other aspects of shared life-styles are determinants of blood pressure level. But the correlation between twin and spouse, while significant, was no higher than that between twin and spouse of co-twin or that between the twins' unrelated spouses, suggesting that resemblance of husband and wife reflects assortative mating rather than environmental covariance.

Kinships of MZ twin parents permit direct tests of maternal influences on human behavior. Kinships of twin mothers can be ascertained as frequently as kinships of twin fathers to provide robust comparisons of the resemblance of maternal and paternal half-siblings. And, in contrast to conventional half-sibships resulting from death, divorce, or illegitimacy, MZ half-sibships are of the same age and size and can be matched for maternal age and parity. An application of the design to verbal IQ measures (Rose, Boughman et al., 1980) suggests that maternal effects may significantly contribute to familial resemblance in verbal intelligence.

Comparisons of the sons and daughters within maternal and paternal half-sibships permit a direct test of X-linkage. Sons within maternal half-sibships inherit their single X chromosome from MZ twin sisters, in contrast to male half-siblings in paternal kinships, whose mothers are genetically unrelated. Such comparisons disconfirm X-linkage as an interpretation of gender differences in perceptual speed ability (Rose, Miller, Dumont-Driscoll & Evans, 1979).

In short, kinships of twin parents permit novel tests of genetic and environmental parameters important in psychosocial epidemiology. The model has been extended to include kinships of DZ twin parents (Haley & Last, 1981), and efforts are under way in Finland (Kaprio, Rose et al., 1987) and the United States (Lykken et al., 1990) to ascertain large and representative samples of kinships of twin parents for research use.

An analysis of the offspring of MZ twins discordant for schizophrenia (Fischer, 1971) illustrates application of these twin-family methods to psychopathology. Fischer reported that risk for schizophrenic outcome for offspring of schizophrenic MZ twin parents was no higher than that found for offspring of the symptom-free co-twins. Equivalent in genetic risk, the offspring differ in exposure to deviant parenting. Fischer's sample was too small to permit rigorous evaluation of the method, but Gottesman and Bertelsen (1989) have reported an 18-year follow-up of Fischer's sample with more compelling results. The (age-corrected) morbid risks for schizophrenia in the offspring of schizophrenic MZ twin parents and in children of their normal co-twins were virtually identical (16.8% and 17.4%); in contrast, risks for offspring of 41 discordant DZ twin pairs in the Danish registry were 17.4% among children of the schizophrenic twin parent, but only 2.1% for children of the normal co-twin. These provocative results from the

Danish twin registry are consistent with a recent report from Norway (Kringlen & Cramer, 1989), based on 73 adult offspring of 22 MZ pairs discordant for schizophenia; prevalence of schizophenia and schizophrenia-like cases did not significantly differ between offspring of schizophrenic twins and those of normal co-twins. Similar results have been reported for manic-depressive disorder: morbid risk is elevated and equivalent in children of affected and normal MZ twin parents, but significantly lower for children of the normal co-twins of discordant DZ twin pairs (Bertelsen & Gottesman, 1986).

Twin Research Designs to Estimate Environmental Effects

Complementing twin studies designed to estimate genetic effects are several research designs that use twins to evaluate environmental sources of variance in behavior and disease. Co-town control studies contrast monozygotic co-twins who differ in environmental experience; because all twin pairs are genetically identical, their intrapair differences permit incisive appraisal of effects of experience specific to one co-twin. Pairwise comparisons, on the other hand, contrast interpair differences among MZ pairs who systematically differ in some dimension of experience; pairs are monozygotic, so again, genetic confounds are eliminated.

Co-twin Control Studies to Evaluate SE

Intrapair comparisons of MZ co-twins who differ in specific experience (*SE*) yield robust tests of effects of that experience. The environmental difference may be experimentally created, as in the work of Miller et al. (1977), who evaluated the therapeutic effectiveness of vitamin C, or the difference may occur as a natural function of varying lifestyles of adult twins, as in comparisons of smoking-discordant co-twins. In either case, the design offers efficiency and power (Christian, 1981), illustrated in the double-blind study of vitamin C by Miller et al. (1977). Subjects were school-aged MZ twins, providing perfect matching on genetic variation in susceptibility and metabolic response; the homogeneous environments of identical twin children closely matched for viral exposure as well. As a consequence, efficiency of the design was, for some variables, 14 times that realized in a population study of genetically unrelated children.

The 1965 WHO report on the use of twins in epidemiological research reviewed the effects of cigarette smoking on mortality and morbidity of MZ twins discordant for smoking. That report was followed (Cederlöf, 1971) by an international symposium that reviewed co-twin control studies testing the relationship of smoking to cardiovascular and pulmonary symptoms. The WHO report recommended an evaluation of potential bias in the use of twins through a comparison of prevalence rates of twins with singletons; the co-twin control comparison then contrasted discordant co-twins to determine whether morbidity was elevated among co-twins exposed to the risk factor. A strong association between smoking and symptoms of chronic bronchitis provides one illustration; that association was evident in the clinical evaluation of smoking-discordant twins in both Sweden (Cederlöf, et al., 1966) and Denmark (Hauge et al., 1970). The most recent report (Kaprio & Koskenvuo, 1989), based on a 12-year prospective study of smoking-discordant Finnish twins, yields conclusive evidence of the health risks caused by smoking. Relative risk of death was 13 times greater for the smoking twin in this population-based sample of discordant MZ twin pairs! These data resolve the central issue in the association of smoking and disease by showing that the association is not a consequence of genetic confounds—a "third variable" effect devoid of causal association. Identical genotypes, differentially exposed to risk, exhibit very different outcomes.

MZ co-twin control designs continue to find new applications. Resnick (1989) advocates sampling discordant MZ twins in neuroimaging research on neuropsychiatric illness; she suggests that co-twin controls provide a sensitive approach to identify and follow the progression of brain abnormalities with neuroimaging techniques in schizophrenia, Alzheimer's and Huntington's diseases, and Tourette syndrome.

MZ Twins Reared Apart and Estimation of CE

To assess effects of shared experience, a second variation on the conventional twin comparison can be used: contrasts of MZ twin pairs who systematically differ in some trait-relevant experience.

The most celebrated of these studies compares MZ twins separated in early life (MZAs) with matched pairs who cohabit until late adolescence or early adulthood (MZTs). This "experiment of nature" was first reported 65 years ago by Muller (1925), who described a pair of MZ twin sisters separated at two weeks due to maternal illness and reared apart until age 18. These sisters were later included in the sample of 19 MZAs studied collaboratively at the University of Chicago (Newman et al., 1937). Publicity surrounding the 1979 reunion of a pair of MZA twin brothers in Ohio has facilitated ascertainment of others and prompted a research team at the University of Minnesota to initiate a contemporary study of MZAs that promises to be the most extensive yet attempted. Some results of that effort have now been reported (Bouchard et al., 1990; Tellegen et al., 1988). (See the chapter by Segal et al. in this volume.)

Selection biases and limited sampling constrain inferences drawn from studies of separated twins, but careful study of MZAs has heuristic value. The report (Juel-Nielsen, 1980) of a 25-year follow-up of a dozen Danish MZAs documents that point. Nine of the 11 deaths recorded at follow-up were due to malignant neoplasms or diseases of the heart. Patterns of concordance for the two categories of disease were very different. No pair was concordant for cancer on 25-year follow-up. Six of the 12 MZA pairs were discordant, and cancer had caused the death of 4 of the 6 affected individuals. Arteriosclerosis was the cause of death of 5 members of the cohort. Three pairs were concordant for arteriosclerotic heart diseases, and another pair was concordant for arterial hypertension. In the remaining case, a twin died at age 69 after a coronary occlusion a decade earlier; her co-twin was alive upon follow-up, aged 73, but suffering slight cardiovascular symptoms. Data from this small, selective sample are illustrative, rather than definitive, but the findings are congruent with others that suggest substantial genetic influences on cardiovascular disease against a predominantly environmental interpretation of cancer.

Assessing CE with MZ Twins Categorized by Age-at-Separation

Although of both popular and scientific interest, studies of birth-separated MZ twins will be invariably of limited size and uncertain ascertainment (Rose, 1982). But the logic of contrasting MZ twin pairs who differ in shared experience need not be so constrained. Rather than narrowly focusing on twins separated at birth, shared experience can be construed as a continuous variable; by measuring age-at-separation (and frequency of subsequent social interaction) we can (crudely) categorize a registry of adult MZ twin pairs by indices of shared environment and can assess their intrapair resemblance. Early results with this approach (Kaprio et al., 1990; Rose & Kaprio, 1988; Tambs et al., 1985) reveal modest, but significant, effects of shared experience on neuroticism, patterns of alcohol use, and adult IQ. Twins in more frequent contact and twins who cohabit longer into adolescence and early adulthood are more alike. Modeling the effect of social contact with robust maximum likelihood techniques (Rose et al., 1990) reveals that frequency of contact makes a significant contribution to twins' similarity in their consumption of alcohol. This pairwise MZ twin design may offer flexible methods for studying environmental factors in behavioral disorder: comparisons of MZ twin pairs whose social experiences differ can assess the impact of diverse environmental differences, from marital or employment status to urban/rural residency. As one example, more than 5,000 adult Finnish twin pairs were linked to Finland's nationwide registry of hospitalizations over a decade's time, and concordances for diagnoses of alcoholism were compared in MZ pairs residing in the same province to MZ pairs living in different provinces; concordance was significantly higher in pairs living in the same province (Romanov et al., 1990). Similar results were found for incidence of psychosis, as accumulated from Finland's nationwide registry of hospital discharges during 1972–1985 (Romanov et al., 1989).

Placental Differences among MZ Twin Pairs

A final twin study comparison yields an assessment of developmental effects of environmental differences associated with placental variation. MZ twins may arise from separation of early blastomeres, so that each developing fetus has its own amnion and chorion with placentation similar to that of DZ twins. More commonly, however, MZ twinning occurs later, through duplication of the inner cell mass, and the twins develop within a single chorion and share a common placenta in which some vessels join via anamestomosis (Bulmer, 1970). Population surveys of newborn Caucasian twins suggest that about one-third of MZs arise from dichorionic (DC) placentation (Corney, 1975). Thus, placental variation among MZ twins provides another pairwise twin comparison of interest in psychosocial epidemiology. Because all pairs are genetically identical, systematic differences between monochorionic (MC) and dichorionic MZ twins must reflect effects of variation in the timing of embryological division and/or vascular communication in utero. Significant placentation effects have been reported for birth weight (Corney, 1978), dermatoglyphics (Reed at al., 1978), cord blood cholesterol (Corey et al., 1976), and 7-year IQ test scores (Melnick et al., 1978). A study of the Block Design subtest of the Wechsler Scales among adult MZ twins of known placentation (Rose, Uchida & Christian, 1981) found DC-MZs no more alike than DZ twins. Dichorionic MZ twins had significantly greater intrapair variation than matched MC-MZ pairs. Such results suggest lasting effects from placental variation in adult MZ twins and encourage its further study.

Future Directions

More than a century after its introduction, the twin method remains a basic tool in epidemiological research, and investigators continue to develop new applications for it. To cite a single example, much recent research suggests that, in addition to conventional applications to issues of etiology, twin data can contribute to questions of nosology. Several investigators (Shields & Gottesman, 1972; Shields & Slater, 1966; Torgersen, 1978) have illustrated that efforts to maximize MZ/DZ concordance ratios have provided insights into the biological utility of alternative classifications for psychiatric disorder. The most recent of these efforts has provided evidence especially significant for diagnostic classification of the schizophrenias. In one study, raters who were blind to zygosity and the psychiatric status of co-twins applied different sets of operational criteria for the diagnosis of schizophrenia to 55 twin pairs, ascertained from the Maudsley Registry. Highest heritabilities were achieved when the Research Diagnostic Criteria for schizophrenia were used (McGuffin et al., 1984); alternative diagnostic criteria proved less useful. In a related study the case histories from 151 pairs of MZ twins, pooled from earlier twin studies of schizophenia, were rated for positive and negative symptoms of the disorder; concordance was significantly higher when MZ probands had more negative symptoms (Dworkin & Lenzenweger, 1984). These novel applications of twin data have implications for resolving heterogeneity in the schizophrenias; such research well illustrates the continuing utility of twin study methods in psychosocial epidemiology.

Acknowledgments

Preparation of this chapter was supported by National Institute on Alcohol Abuse and Alcoholism grants AA-06232, AA-07611, and AA-08315.

References

Bertelsen, A. & Gottesman, I. I. (1986) Offspring of twin pairs discordant for psychiatric illness. Presented at the 5th International Congress on Twin Studies, Amsterdam.

Bouchard, T. J., Jr., Lykken, D. T., McGue, M., Segal, N. L. & Tellegen, A. (1990) Sources of human psychological differences: The Minnesota Study of Twins Reared Apart. *Science* 250:223–228.

Bulmer, M. G. (1970) *The Biology of Twinning in Man*. Oxford: Clarendon.

Cederlöf, R. (Ed.) (1971) Twin registries in the study of chronic disease with particular reference to the relation of smoking to cardiovascular and pulmonary diseases. *Acta Medica Scandinavica* 523 (Suppl.):1–40.

Cederlöf, R., Friberg, L., Johnson, E. & Kaij, L. (1966) Respiratory symptoms and "angina pectoris" in twins with reference to smoking habits: An epidemiological study with mailed questionnaire. *Archives of Environmental Health* 13:706–737.

Christian, J. C. (1979) Testing twin means and estimating genetic variance: Basic methodology for the analysis of quantitative twin data. *Acta Geneticae Medicae et Gemellologiae* 28:35–40.

Christian, J. C. (1981) Use of twins to study environmental effects. *Environmental Health Perspectives* 41:103–106.

Corey, L. A., Kang, K. W., Christian, J. C., Norton, J. A., Jr., Harris, R. E. & Nance, W. E. (1976) Effects of chorion type on variation in cord blood cholesterol of monozygotic twins. *American Journal of Human Genetics* 28:433–441.

Corey, G. (1975) Placentation. In *Human Multiple Reproduction*, ed. I. MacGillivray, P.P.S. Nylander & G. Corney, pp. 40–76. London: W. B. Saunders.

Corney, G. (1978) Twin placentation and some effects on twins of known zygosity. In *Twin Research: Biology and Epidemiology*, ed. W. E. Nance, G. Allan & P. Parisi, pp. 9–16. New York: A. R. Liss.

Dworkin, R. H. & Lenzenweger, M. F. (1984) Symptoms and the genetics of schizophrenia: Implications for diagnosis. *American Journal of Psychiatry* 141:1541–1546.

Eaves, L. J. (1978) Twins as a basis for the causal analysis of human personality. In *Twin Research: Psychology and Methodology*, ed. W. E. Nance, G. Allan & P. Parisi, pp. 151–174. New York: A. R. Liss.

Fischer, M. (1971) Psychoses in the offspring of schizophrenic monozygotic twins and their normal co-twins. *British Journal of Psychiatry* 18:43–52.

Fischer, M., Harvald, B. & Hauge, M. (1969) A Danish twin study of schizophrenia. *British Journal of Psychiatry* 115:981–990.

Gottesman, I. I. & Bertelsen, A. (1989) Confirming unexpressed genotypes for schizophrenia. *Archives of General Psychiatry* 46:867–872.

Gottesman, I. I. & Shields, J. (1966) Schizophrenia in twins: 16 years' consecutive admissions to a psychiatric clinic. *British Journal of Psychiatry* 112:809–818.

Gottesman, I. I. & Shields, J. (1972) *Schizophrenia and Genetics: A Twin Study Vantage Point*. New York: Academic Press.

Haley, C. S. & Last, K. (1981) The advantages of analyzing human variation using twins and twin half-sibs and cousins. *Heredity* 46:227–238.

Hauge, M., Harvald, B. & Reid, D. D. (1970) A twin study of the influence of smoking on morbidity and mortality. *Acta Geneticae Medicae et Gemellologiae* 19:335–336.

Heath, A. C., Kendler, K. S., Eaves, L. J. & Markell, D. (1985) The resolution of cultural and biological inheritance: Informativeness of different relationships. *Behavior Genetics* 15:439–465.

Inouye, E. (1963) Similarity and dissimilarity of schizophrenia in twins. In *Proceedings, Third World Congress on Psychiatry*, vol. 1, pp. 524–530. Montreal: University of Toronto Press.

Jackson, D. D. (1960) A critique of the literature on the genetics of schizophrenia. In *The Etiology of Schizophrenia*, ed. D. D. Jackson, pp. 37–87. New York: Basic Books.

Jinks, J. L. & Fulker, D. W. (1970) Comparison of the biometrical genetical, MAVA, and classical approaches to the analysis of human behavior. *Psychological Bulletin* 73:311–349.

Juel-Nielsen, N. (1980) *Individual and Environment: Monozygotic Twins Reared Apart*, rev. ed. New York: International Universities Press.

Kaprio, J. & Koskenvuo, M. (1989) Twins, smoking and mortality: A twelve-year prospective study of smoking-discordant twin pairs. *Social Science & Medicine* 12:1083–1089.

Kaprio, J., Koskenvuo, M., Langinvainio, H., Romanov, K., Sarna, S. & Rose, R. J. (1987) Genetic influences on use and abuse of alcohol: A study of 5638 adult Finnish twin brothers. *Alcoholism: Clinical & Experimental Research* 11:349–356.

Kaprio, J., Koskenvuo, M. & Rose, R. J. (1990) Change in cohabitation and intra-pair similarity of MZ co-twins for alcohol use, extraversion, and neuroticism. *Behavior Genetics* 20:265–276.

Kaprio, J., Koskenvuo, M. & Rose, R. J. (in press) Population-based twin registries in the Nordic countries: Illustrative applications in genetic epidemiology and behavioral genetics. *Acta Geneticae Medicae et Gemellologiae*.

Kaprio, J., Rose, R. J., Sarna, S., Langinvainio, H., Koskenvuo, M., Rita, H. & Heikkilä, K. (1987) Design and sampling considerations, response rates, and representativeness in a Finnish twin family study. *Acta Geneticae Medicae et Gemellologiae* 36:79–93.

Kendler, K. S. & Robinette, C. D. (1983) Schizophrenia in the National Academy of Sciences-National Research Council Twin Registry: A 16-year update. *American Journal of Psychiatry* 140:1551–1563.

Kringlen, E. (1967) *Heredity and Environment in the Functional Psychoses: An Epidemiological Clinical Twin Study*. London: Heinemann.

Kringlen, E. & Cramer, G. (1989) Offspring of monozygotic twins discordant for schizophrenia. *Archives of General Psychiatry* 46:873–877.

Loehlin, J. C. (1989) Partitioning environmental and genetic contributions to behavioral development. *American Psychologist* 44:1285–1292.

Loehlin, J. C. & Nichols, R. C. (1976) *Heredity, Environment, and Personality*. Austin: University of Texas Press.

Lykken, D. T., Bouchard, T. J., Jr., McGue, M. & Tellegen, A. (1990) The Minnesota Twin Family Registry: Some initial findings. *Acta Geneticae Medicae et Gemellologiae* 39:35–70.

Lykken, D. T., McGue, M. & Tellegen, A. (1987) Recruitment bias in twin research: The rule of two-thirds reconsidered. *Behavior Genetics* 17:343–362.

Lykken, D. T., Tellegen, A. & DuRubeis, R. (1978) Volunteer bias in twin research: The rule of two-thirds. *Social Biology* 25:1–9.

Lytton, H. (1977) Do parents create or respond to differences in twins? *Developmental Psychology* 13:456–459.

Martin, N. G. & Wilson, S. R. (1982) Bias in the estimation of heritability from truncated samples of twins. *Behavior Genetics* 12:467–472.

Matheny, A. P., Jr. (1979) Appraisal of parental bias in twin studies. Ascribed zygosity and IQ differences in twins. *Acta Geneticae Medicae et Gemellologiae* 28:155–160.

Matheny, A. P., Jr., Wilson, R. S. & Dolan, A. B. (1976) Relations between twins' similarity of appearance and behavioral similarity: Testing an assumption. *Behavior Genetics* 6:343–351.

McGuffin, P., Farmer, A. E., Gottesman, I. I., Murray, R. M. & Reveley, A. M. (1984) Twin concordance for operationally defined schizophrenia:

phrenia: Confirmation of familiality and heritability. *Archives of General Psychiatry* 41:541–545.

Melnick, M., Myrianthopoulos, N. C. & Christian, J. C. (1978) The effects of chorion type on variation in IQ in the NCPP twin population. *American Journal of Human Genetics* 30:425–433.

Miller, J. Z., Nance, W. E., Norton, J. A., Wolen, R. L., Griffith, R. S. & Rose, R. J. (1977) Therapeutic effect of vitamin C: A co-twin control study. *Journal of the American Medical Association* 237:248–251.

Muller, H. J. (1925) Mental traits and heredity: The extent to which mental traits are independent of heredity as tested in a case of identical twins reared apart. *Journal of Heredity* 16:433–437.

Newman, H. H., Freeman, F. N. & Holzinger, K. J. (1937) *Twins: A Study of Heredity and Environment*. Chicago: University of Chicago Press.

Phillips, K., Fulker, D. W. & Rose, R. J. (1987) Path analysis of seven fear factors in adult twin and sibling pairs and their parents. *Genetic Epidemiology* 4:345–355.

Plomin, R., Willerman, L. & Loehlin, J. C. (1976) Resemblance in appearance and the equal environments assumption in twin studies of personality traits. *Behavior Genetics* 6:43–52.

Pollin, W., Allen, M. G., Hoffer, A., Stabenau, J. R. & Hrubec, Z. (1969) Psychopathology in 15,909 pairs of veteran twins: Evidence for a genetic factor in the pathogenesis of schizophrenia and its relative absence in psychoneurosis. *American Journal of Psychiatry* 126:597–610.

Reed, T., Uchida, I. A., Norton, J. A., Jr. & Christian, J. C. (1978) Comparisons of dermatoglyphic patterns in monochorionic and dichorionic monozygotic twins. *American Journal of Human Genetics* 30:383–391.

Resnick, S. M. (1989) The application of genetic strategies to neuroimaging studies of neuropsychiatric illness. *Behavior Genetics* 19:773 (Abstr.).

Romanov, K., Kaprio, J., Koskenvuo, M., Rose, R. J., Sarna, S. & Heikkilä, K. (1990) Genetics of alcoholism: The effect of migration among male twins. Presented at meetings of the International Society for Biological Research on Alcoholism, Toronto.

Romanov, K., Koskenvuo, M., Kaprio, J. & Heikkilä, K. (1989) Selection bias in twin studies. *Acta Geneticae Medicae et Gemellologiae* 38:138 (Abstr.).

Rose, R. J. (1982) Separated twins: Data and their limits. *Science* 213:959–960.

Rose, R. J., Boughman, J. A., Corey, L. A., Nance, W. E., Christian, J. C. & Kang, K. W. (1980) Data from kinships of monozygotic twins indicate maternal effects on verbal intelligence. *Nature* 283:375–377.

Rose, R. J. & Ditto, W. B. (1983) A developmental-genetic analysis of common fears from early adolescence to early adulthood. *Child Development* 54:361–368.

Rose, R. J., Fulker, D. W., Miller, J. Z., Grim, C. E. & Christian, J. C. (1980) Heritability of systolic blood pressure: Analysis of variance in MZ twin parents and their children. *Acta Geneticae Medicae et Gemellologiae* 29:143–149.

Rose, R. J., Harris, E. L., Christian, J. C. & Nance, W. E. (1979) Genetic variance in nonverbal intelligence: Data from the kinships of identical twins. *Science* 205:1153–1155.

Rose, R. J. & Kaprio, J. (1988) Frequency of social contact and intrapair resemblance of adult monozygotic cotwins. *Behavior Genetics* 18:309–328.

Rose, R. J., Kaprio, J., Williams, C. J., Viken, R. & Obremski, K. (1990) Social contact and sibling similarity: Facts, issues, and red herrings. *Behavior Genetics* 20:763–778.

Rose, R. J., Koskenvuo, M., Kaprio, J., Sarna, S. & Langinvainio, H. (1988) Shared genes, shared experiences, and similarity of personality. *Journal of Personality & Social Psychology* 54:161–171.

Rose, R. J., Miller, J. Z., Dumont-Driscoll, M. & Evans, M. M. (1979) Twin-family studies of perceptual speed ability. *Behavior Genetics* 9:71–86.

Rose, R. J., Miller, J. Z., Grim, C. E. & Christian, J. C. (1979). Aggregation of blood pressure in the families of identical twins. *American Journal of Epidemiology* 109:503–511.

Rose, R. J., Miller, J. Z., Pogue-Geile, M. F. & Cardwell, G. F. (1981) *Twin-Family Studies on Common Fears and Phobias. Twin Research 3: Intelligence, Personality and Development*. New York: A. R. Liss, 169–174.

Rose, R. J., Uchida, I. A. & Christian, J. C. (1981) *Placentation Effects on Cognitive Resemblance of Adult Monozygotes. Twin Research 3: Intelligence, Personality and Development*. New York: A. R. Liss, 35–41.

Rosenthal, D. (1960) Confusion of identity and the frequency of schizophrenia in twins. *Archives of General Psychiatry* 3:297–304.

Scarr, S. (1968) Environmental bias in twin studies. *Eugenics Quarterly* 15:34–40.

Shields, J. & Gottesman, I. I. (1972) Cross-national diagnosis of schizophrenia in twins. *Archives of General Psychiatry* 27:725–730.

Shields, J. & Slater, E. (1966) La similarit du diagnostic chez les jumeaux et le problme de la spcificit biologique dans les nvroses et les troubles de la personalit. *L'Evolution Psychiatrique* 31:441–451. [Original English version in *Man, Mind, and Heredity: Selected Papers of Eliot Slater on Psychiatry and Genetics*, ed. J. Shields & I. I. Gottesman, pp. 552–557. (1971) Baltimore: Johns Hopkins University Press.]

Tambs, K., Sundet, J. M. & Berg, K. (1985) Cotwin closeness in zygotic and dizygotic twins: A biasing factor in IQ heritability analysis? *Acta Geneticae Medicae et Gemellologiae* 34:33–39.

Tellegen, A., Lykken D. T., Bouchard, T. F., Jr., Wilcox, K. J., Segal, N. L. & Rich, S. (1988) Personality similarity in twins reared apart and together. *Journal of Personality & Social Psychology* 54:1031–1039.

Tienari, P. (1968) Schizophrenia in monozygotic male twins. In *The Transmission of Schizophrenia*, ed. D. Rosenthal & S. S. Kety. Oxford: Pergamon Press.

Tischler, P. V., Lewitter, F. I., Rosner, B. & Speizer, F. E. (1987) Genetic and environmental control of blood pressure in twins and their family members. *Acta Geneticae Medicae et Gemellologiae* 36:455–466.

Torgersen, S. (1978) The contribution of twin studies to psychiatric nosology. In *Twin Research: Psychology and Methodology*, ed. W. E. Nance, G. Allan & P. Parisi. New York: A. R. Liss.

World Health Organization (1966) The use of twins in epidemiological studies: Report of the WHO meeting of investigators in twin studies. *Acta Geneticae Medicae et Gemellologiae* 15:109–128.

Chapter 3
Adoption Studies in Psychosocial Epidemiology

REMI J. CADORET

In the study of psychosocial epidemiology in psychiatry, adoptees and various separation paradigms provide a most powerful tool to separate and characterize genetic, environmental, and genetic-environmental interactional etiologies. The 1960s saw an increased interest in the study of adoptees for evidence of genetic transmission of psychopathology. However, it has only been in more recent years that adoption studies have shown that environmental factors and gene-environment interaction are important in psychopathology (Cadoret, 1986). In the late 1980s, researchers began to focus on different types of environmental factors to explain within-family individual differences. For example, nonshared environmental differences between children in the same family have been hypothesized to account for a large proportion of the variance for psychopathology and even for normal personality variants and cognitive abilities (Plomin & Daniels, 1987). Adoption studies also provide a way to study nonshared environment directly, by sampling one adoptee from a number of adoptive families, for example. In such a study between-family differences offer a measure of nonshared environment. Other interactions of the environment with genotypes, such as three different kinds of genotype-environment correlations proposed by Scarr and McCartney (1983), could be studied to advantage with a separation paradigm since it would permit a cleaner definition of genotype and environment and the effect of the interaction.

There are other current issues in psychiatry that adoption designs are able to clarify. In nosology, spectrum conditions have been proposed to explain behavioral abnormalities found in the relatives of patients with diseases like schizophrenia. Examining adopted-away offspring of individuals with core conditions such as unequivocal schizophrenia would be one way to define a schizophrenic spectrum, and this approach has been used by Kendler et al. (1981a; 1981b) to show that schizotypal conditions and delusional disorder are related to schizophrenia. Such adoption designs could also determine which, if any, environmental factors might be responsible for affecting the outcome of spectrum conditions, provided environmental factors were assessed.

With the advent of readily available genetic probes in molecular genetics many studies have been launched to detect evidence of genetic linkage to psychiatric illness. In this strategy mathematical models are fit to family trees containing some patients with psychiatric illness. If a patient's illness is due to environmental factors and not to a genetic effect, then inclusion in the mathematical model of such an individual (called a phenocopy) would decrease the power to detect linkage. Adoption studies can be used to determine what environmental conditions are associated with phenocopies, and to make appropriate adjustments in weighting the psychiatric outcome of individuals who have been included in linkage models (Cadoret & Wesner, 1990).

This chapter will present the separation paradigm and indicate practical ways to carry out such a design.

The Separation Paradigm

The General Paradigm and Its Rationale

The purpose of the separation, or adoption, paradigm is to unconfound factors that in ordinary family life would be confounded, as when natural-born children are raised by their biologic parents. Genetic background factors and postnatal environmental factors are unconfounded by physical separation of the infant at an early age (ideally at birth) and by placement with nonrelatives. This type of study was first proposed in 1912 (Richardson, 1912–1913) as a way of determining the relative importance of "nature" (heredity) versus "nurture" (environment) in human intelligence. Many Western countries have developed a strong tradition of placing newborns with nonrelatives (other countries, e.g., Japan, frequently adopt to relatives in order to carry on the family name). England, Germany, the Scandinavian countries, the United States, and Canada have made extensive

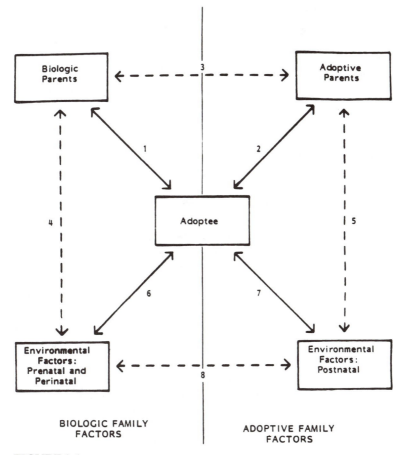

FIGURE 3.1.
The adoption paradigm showing the relationship between
hereditary and environmental factors.

use of adoption by nonrelatives and it is this practice that can help
to separate genetic and environmental factors.

Figure 3.1 shows the genetic and environmental factors that
impinge upon an adoptee. The vertical line depicts the role of the
separation in unconfounding biologic parental factors from adoptive family factors. If the placement of the adoptee is with a
nonrelative and placement is not biased by factors in the biologic
family background, then relationships 3 and 8 in Figure 3.1 will be

nonexistent and need not be considered in assessing 1, 2, 6, and 7—the genetic and environmental factors. One major criticism of adoption designs has been the possibility that adoption practices might lead to matching of biologic with adoptive family characteristics (Lewontin et al., 1984). An obvious example is the practice of agencies to match children with parents on the basis of physical characteristics such as hair and eye color. Matching on psychosocial characteristics, known generally as "selective placement," could lead to erroneous conclusions about the importance of factors 1, 2, 6, and 7. Selective placement is a very realistic concern since adoptees are not assigned at random to adoptive homes.

In the diagram, 1 can be considered a measure of the genetic factors contributed to the adoptee by his/her biologic parents. Relationship 2 can be conceived as a measure of the quality of adoptive parent interaction with the adoptee, a factor that could also reflect the reaction of the adoptive parent to the adoptee's behavior (e.g., difficult temperament). Parents, whether biologic or adoptive, determine the quality of many environmental factors. These relationships are indicated by factors 4 and 5, and these correlated environmental factors in turn affect the adoptee (and vice versa) through relationships 6 and 7. By identifying and measuring the factors appearing in the parent and environmental boxes, the effects of 1, 6, and 7 can be determined. The diagram indicates that adoptee behavior is a multivariate problem; it requires statistical analyses that simultaneously account for the combined effect of a number of variables and determine the significance of the contribution of separate factors to the whole. In part, the selective placement problem alluded to above can be controlled statistically by including in multivariate models the psychosocial or other factors presumed to be involved in the selective placement. More modern adoption studies have used sophisticated multivariate analyses to control for factors such as selective placement by use of log-linear modeling (Cadoret, 1986; Cadoret, O'Gorman, Heywood & Troughton, 1985; Cadoret, O'Gorman, Troughton & Heywood, 1985). In none of the studies cited was evidence for selective placement found. Additional multivariate techniques can be used to control genetic factors while testing for environmental effects (Cadoret & Cain, 1981). Such studies have been carried out by Cloninger et al. in alcoholism and criminality (Bohman, et al. 1982; Cloninger, et al. 1982; Sigvardsson, et al. 1982) using discriminant function as a basis for the analysis, and by Cadoret et al. in antisocial behavior (Cadoret, O'Gorman, Troughton, & Heywood, 1985), alcoholism (Cadoret et al., 1980; Cadoret, O'Gorman, Troughton & Heywood, 1985), and depression (Cadoret, O'Gorman, Heywood & Troughton, 1985) using log-linear modeling. An additional multivariate approach proposed by Cloninger et al. (1979) is the use of path analysis.

Specific Types of Separation Paradigms

Types of separation designs can be classified by two characteristics: (1) the completeness of information gathered on the adoptee, on the biologic, and on the adoptive parents and (2) whether the adoptee or a biologic parent is the starting point for the study.

By far the most common "complete" adoption design is one that starts with information about the biologic parents and then determines the outcome of the adoptee (as well as examining the environmental conditions provided by the adoptive family). This has been termed by Rosenthal (1970) as the "adoptees' study method." In contrast to this strategy, when the starting point is the adoptee and the follow-up seeks information about the biologic and adoptive parents, Rosenthal designated the design the "adoptees' family method."

Both these designs are "complete" in the sense that information is available about biologic and adoptive families, as well as about the adoptee. The starting point determines to some degree the interpretation of results. Both designs can demonstrate genetic factors, but environmental factors could be confounded with selection of the proband adoptee in the adoptees' family method design; for example, when psychiatrically ill adoptees are selected as probands as a starting point for the study, environmental factors such as lower socioeconomic status could have determined whether they sought treatment. However, if an entire population of treated individuals (as in some of the Danish studies) is used as a starting point, such confounding might not occur.

Comorbidity, or the occurrence of more than one psychiatric condition in a single individual, has been an increasingly important consideration in clinical studies. If adoption designs are used to clarify the sources of comorbidity—whether genetic, environmental, or the result of gene-environment interaction—then the first design, the adoptees' study method, is probably to be preferred since the adoptees in whom comorbid conditions are to be assessed are selected on the basis of biologic parent characteristics and not, as in the second design, from clinical populations in which comorbidity may be a factor leading to increased treatment seeking.

Partial separation designs occur when, in addition to adoptee outcome, information is available for only the adoptive family. For example, it is possible to examine correlations between adoptees and their adoptive parents and compare these correlations with those found between parents and their natural children. This approach has been used, for example, in IQ research (Scarr & Weinberg, 1981; Speer, 1940). Most partial designs use the adoptee as a starting point, and sampling bias is thus a strong possibility (e.g., the sample may contain more severely ill probands, those with comorbid conditions, or those who are exposed to unusual environmental factors that lead to an increased chance of seeking treatment).

Variants and Combinations
of the Basic Separation Design

A high risk paradigm is commonly included in the basic separation design. Probands with a certain condition, whether the adoptee or the biologic parent, usually comprise the entire "experimental" group, and are contrasted with a control group selected because of normalcy or at least lack of the condition characterizing the experimental group. Most reported adoption studies utilize this high risk approach.

Many adoption agencies handle late adoptions wherein children who have lived with biologic parents for months or years are removed from the home (generally for neglect and/or abuse) and adopted into nonrelated families. This provides an opportunity for investigating the effect of early life neglect/abuse upon later behavior. The problem of finding a suitable control or comparison group can be solved in part by taking advantage of the fact that such families often have a number of children among whom are newborns who are also adopted away and can serve as a control.

One variant of the separation design that has not been exploited by research is the study of foster children who, for various reasons, have never been legally adopted. Reasons for nonadoption would have to be considered in evaluating such a study since nonadoptability (because of conduct disorder or physical handicaps, for example) could be confounded with adoptee outcome. However, there are many foster children separated close to birth who might be studied along with their adopted-away counterparts. In this way, the number in a study could be increased. Many children removed from home because of abuse/neglect are never adopted because the courts do not sever parental rights though the child may remain in foster care for years.

Controls and Comparison Groups

In the Rosenthal adoptees' study design, control adoptees for outcome are generally adoptees from a contrasting biologic background (normal or with a different type of psychopathology) who are matched for important controlling variables such as age and sex. In the Rosenthal adoptees' family design, controls for adoptee probands are usually selected for normalcy and may also be matched on a variety of relevant variables such as age, sex, and socioeconomic level.

However, there are a number of other comparison groups available in most adoption designs: natural-born children of adoptive parents and nonadopted-away sibs of adoptees. Use of such groups can help answer questions about genetic or environmental etiology. For example, Goodwin et al. (1977) showed that a group of nonadopted-away daughters of alcoholics had higher depression rates than their adopted-away sibs, a finding suggestive of an environmental effect or a gene-environment interaction occurring in the presence of an alcoholic genetic diathesis.

Determination of Adoptee Outcome

In the Scandinavian countries much use has been made of central registers for hospitalization or treatment of psychiatric illness, alcohol abuse, and incarceration to characterize adoptee outcome in studies that start with proband biologic parents. In some cases (Goodwin et al., 1973; Rosenthal & Kety, 1968) adoptees were personally interviewed by "blinded" interviewers to minimize observer bias.

In North America, where such registries are generally not available, outcome is often measured by an interview of the adoptee that allows the use of a structured instrument. Further information about adoptee behavior can also be gathered from interviews of adoptive parents and other family members. Through these contacts, hospital records, school records, and other external sources can be traced for further information.

Measuring outcome by personal interview generally increases the harvest of psychiatric conditions that often remain untreated, yet may be significant to the individual and relevant to the investigator's hypotheses. For example, in a recent analysis of illicit drug use in an adoptee sample, the present investigator found that of 28 drug-abusing males, only 8 had ever had professional contact to treat the condition.

Determination of Environmental Factors

If adoptive families can be interviewed, a great deal can be learned about the postnatal environment. Much social and demographic information can be elicited in parental interviews to assess such factors as marital problems, separations, and divorce. Structured psychiatric interviews can be given to adoptive parents to detect the presence of important factors such as alcoholism, depression, or psychosis. Access to adoptees and adoptive families to carry out these studies can be made through the adoption agency.

Information is usually available from adoption agency records (which often contain a copy or summary of hospital records) relating to prenatal care and to the delivery and neonatal period. Many mothers who give up their children for adoption do not receive much prenatal care, however, so that records may be scanty regarding health and dietary factors during the pregnancy. Since it is usually not possible to interview biologic parents these environmental factors may have to be overlooked to the detriment of the study. Adoption agencies, by law, are required to demonstrate that the adoptee is adjusting well in the adoptive home before an adoption is finalized. Records of home visits and related material from this postnatal period are therefore usually available from the agency and can be used to assess such factors as parental bonding and early adjustment of the child to the home.

Other information about the early months of adoptee life prior to final adoptive home placement are usually well documented by the adoption agency. This includes such factors as the number of temporary foster homes, the adoptee's age at placement into the permanent adoptive home, and other experiences such as prolonged medical observation of the adoptee for possible developmental problems. These are vicissitudes to which the non-adopted are not exposed and which may affect later life behavior (Cadoret et al., 1990; von Knorring et al., 1982), hence the importance of including such perinatal and postnatal factors in analyzing an adoption study.

Adoptive parents are carefully screened by adoption agencies for their parenting potential. Thus it is less likely that severe social or psychiatric problems will be present in such parents. This truncation of the range of psychosocial variables in adoptive homes can lead to underestimation of the importance of the environment because of the weakening of correlations based upon the limited range of a variable. Adoptive families are often not representative of the general population in other respects. One of these is age. Most agencies adopt newborns only to "childless" couples. The age of such parents is generally older since it takes a number of years to demonstrate failure to conceive and bear children.

In spite of careful adoptive family selection, a significant amount of psychiatric and social problems do turn up in adoptive families. This investigator has found evidence for significant psychiatric or behavioral problems in approximately one-third of adoptive homes (Cadoret, O'Gorman, Troughton & Heywood, 1985). However, some social problems such as divorce are markedly lower than population values (Cadoret, O'Gorman, Troughton & Heywood, 1985), reflecting the success of agencies in selecting stable families. Thus the evidence suggests that while adoptive families are not representative of all families in the population, there is still a range of psychosocial variation in a number of variables that can be examined for their effect upon adoptee outcome.

Determination of Genetic Factors

Assessment of the phenotype of the biologic parents is one of the most important aspects of the adoption study. Yet, it can be one of the weakest. In adoption studies done in the Scandinavian countries, central registries have been available for determination of psychiatric conditions requiring hospitalization (such as depression or schizophrenia), for incarceration for criminal activity, and for treatment of alcoholism. Use of such registries has resulted in well-documented parental behaviors that sometimes match current diagnostic criteria for psychiatric diseases. However, in other parts of the world, such as North America, documentation of biologic parental conditions cannot be obtained from central registries. Sources of information are usually limited to adoption agency records, which to a greater or lesser degree try to characterize the qualities of the biologic parent. Since most agencies deal mainly with the biologic mother, there is usually more information in the record about the maternal side of the biologic parentage. In earlier years before it was necessary to obtain release of the child from the father, information about the paternal side could be very scanty indeed.

Other means of characterizing the biologic factors are available. At least two North American studies have used institutional records indicating that a child was given up for adoption. One of these is the classic study of Heston (1966), who followed adopted-away children of schizophrenic mothers who had given birth while institutionalized. The other study is that of Crowe (1974), who found a number of adopted-away offspring of convicted female felons by examining prison records. Recently a prospective adoption study in Colorado (Plomin & DeFries, 1985) recruited biologic parents who were giving up their infant for adoption. This enabled the investigators to personally interview and test the biologic parents to establish phenotypes, and so on. In an environment where both biologic parents must sign away their parental rights, prospective studies like the Colorado project would be feasible, and should be more widely used to answer genetic/environmental hypotheses about childhood hyperactivity, conduct disorders, and other childhood psychiatric problems in which 10 to 12 years of prospective observation should yield answers.

Although North American equivalents of the Scandinavian central registries do not exist, there are enough scattered sources of information about biologic parents to warrant a record linkage to create a registry of individuals with various psychiatric diagnoses who have given up children for adoption. The author is creating such a registry in the state of Iowa. The procedure will be described here since many of its features could be copied and registries constructed in other areas of the United States or perhaps in other countries where Western-style adoption is practiced.

The starting point for the registry is to ascertain children given up at birth and successfully adopted by nonrelatives. This is accomplished by searching records of a number of adoption agencies, from which names and identifying information about biologic parents are available. The next step is to determine which of these biologic parents have been treated for a psychiatric condition; in Iowa we searched records of the state's four mental health institutes and several other large tertiary care hospitals for matches to the biologic parent names. When hospital records were found they were reproduced with parent names and other parental identifying data deleted. However, because of a coding technique used by the adoption agencies in identifying biologic parents, we were able to match a biologic parent to his/her adopted-away child without identifying the parent by name in our data set. This provision satisfied current state and federal law with regard to confidentiality. Having the hospital records has meant that diagnoses of biologic parents will be better documented and that it should be possible to apply modern psychiatric criteria for diagnosis of mental conditions.

A number of other public-maintained records will also be used to determine biologic parent diagnosis. The state maintains prison records going back a number of years. These records usually have a very complete social history so that they are appropriate aids in determining conditions such as antisocial personality and substance abuse. Upon entrance to the prison system, inmates are usually tested to determine their placement within the correctional system, and this further information, such as IQ or personality tests, is often available. In Iowa, access to these records is through a centralized index or registry, and it is likely that records in other states are organized similarly.

Another source of information is death certificates, which are often computerized. Such records can be used to find people with alcohol-related diagnoses, such as Laennec's cirrhosis, or to determine suicide or other unnatural death that could be related to a psychiatric condition.

Using these sources of information we plan to construct a registry of parents (nameless but identified by a code number that ties them to an adoptee) with a variety of psychiatric conditions that have resulted in institutional contact. This registry can then be used to select proband parents and controls or contrast groups to test genetic hypotheses about the psychiatric conditions. Access to the adoptee and the adoptive family will be accomplished as usual through the adoption agency.

Although the procedures described above are tedious, the use of computers and computerized records can facilitate searches, as well as help organize and keep track of the resulting large data sets.

Summary and Conclusions

Separation designs represent a viable approach to the solution of questions regarding importance of genetic predisposition, environmental causes, and the genetic-environmental interaction in psychiatric etiology. North American investigators have shown that it is possible to perform adoption studies without the use of central registries, and it is hoped that the future will see more of these studies contribute to our understanding of psychosocial epidemiology.

References

Bohman, M., Cloninger, C. R., Sigvardsson, S. & von Knorring, A. L. (1982) Predisposition to petty criminality in Swedish adoptees: I. Genetic and environmental heterogeneity. *Archives of General Psychiatry* 11:1233–1241.

Cadoret, R. J. (1986) Adoption studies: Historical and methodological critique. *Psychiatric Development* 1:45–64.

Cadoret, R. J. & Cain, C. (1981) Environmental and genetic factors in predicting adolescent antisocial behavior in adoptees. *Psychiatric Journal of the University of Ottawa* 6:220–225.

Cadoret, R. J., Cain, C. & Grove, W. (1980) Development of alcoholism in adoptees raised apart from alcoholic biologic relatives. *Archives of General Psychiatry* 37:561–563.

Cadoret, R. J., O'Gorman, T. W., Heywood, E. & Troughton, E. (1985) Genetic and environmental factors in major depression. *Journal of Affect Discorders* 9:155–164.

Cadoret, R. J., O'Gorman, T., Troughton, E. & Heywood, E. (1985) Alcoholism and antisocial personality: Interrelationships, genetic and environmental factors. *Archives of General Psychiatry* 42:161–167.

Cadoret, R. J., Troughton, E., Moreno, L. & Whitters, A. (1990) Early life psychosocial events and adult affective symptoms. In *Strait and Devious Pathways from Childhood to Adulthood*, ed. M. Rutter & L. Robbins, pp. 300–313. Cambridge: Cambridge University Press.

Cadoret, R. J. & Wesner, R. (1990) Use of the adoption paradigm to elucidate the role of genes and environment and their interaction in the genesis of alcoholism. In *Banburgy Report 33: Genetics and Biology of Alcoholism*, ed. C. R. Cloninger & H. Begleiter, 380 pp. Cold Spring Harbor Laboratory Press.

Cloninger, C. R., Rice, J. & Reich, J. (1979) Multifactorial inheritance with cultural transmission and assortive mating: III. Family structure and the analysis of separation experiments. *American Journal of Human Genetics* 31:366–388.

Cloninger, C. R., Sigvardsson, S., Bohman, M. & von Knorring, A. L. (1982) Predisposition to petty criminality in Swedish adoptees: II. Cross-fostering analysis of gene-environment interaction. *Archives of General Psychiatry* 11:1242–1247.

Crowe, R. R. (1974) An adoption study of antisocial personality. *Archives of General Psychiatry* 31:785–791.

Goodwin, D. W., Schulsinger, F. & Hermansen, L. (1973) Alcohol problems in adoptees raised apart from alcoholic biological parents. *Archives of General Psychiatry* 28:238–243.

Goodwin, D. W., Schulsinger, F. & Knop, J. (1977) Psychopathology in adopted and nonadopted daughters of alcoholics. *Archives of General Psychiatry* 34:105–109.

Heston, L. L. (1966) Psychiatric disorders in foster home reared children of schizophrenic mothers. *British Journal of Psychiatry* 112:819–825.

Kendler, K. S., Gruenberg, A. M. & Strauss, J. S. (1981a) An independent analysis of the Copenhagen sample of the Danish adoption study of schizophrenia: II. The relationship between schizotypal personality disorder and schizophrenia. *Archives of General Psychiatry* 38:982–984.

Kendler, K. S., Gruenberg, A. M. & Strauss, J. S. (1981b) An independent analysis of the Copenhagen sample of the Danish adoption study of schizophrenia: III. The relationship between paranoid psychosis (delusional disorder) and schizophrenia. *Archives of General Psychiatry* 38:985–987.

Lewontin, R. C., Rose, S. & Kamin, L. (1984) Not in our genes. *Biology, Ideology, and Human Nature*. New York: Pantheon Books.

Plomin, R. & Daniels, D. (1987) Why are children in the same family so different from one another? *Behavioral and Brain Sciences* 10:1–60.

Plomin, R. & DeFries, J. C. (1985) Origins of individual differences in infancy. In *The Colorado Adoption Project*. New York: Academic Press.

Richardson, L. F. (1912–1913) The measurement of mental "nature" and the study of adopted children. *Eugenics Review* 4:391–394.

Rosenthal, D. (1970) *Genetic Theory and Abnormal Behavior.* New York: McGraw-Hill.

Rosenthal, D. & Kety, S. S. (1968) *Transmission of Schizophrenia.* Oxford: Pergamon Press.

Scarr, S. & McCartney, K. (1983) How people make their own environments: A theory of genotype environment effects. *Child Development* 54:424–435.

Scarr, S. & Weinberg, R. A. (1981) Intellectual similarities within families of both adopted and biological children. In *Race, Social Class, and Individual Differences in I.Q.,* ed. S. Scarr, pp. 319–383. Hillsdale, NJ: L. Erlbaum Associates.

Sigvardsson, S., Cloninger, C. R., Bohman, M. & von Knorring, A. L. (1982) Predisposition to petty criminality in Swedish adoptees: III. Sex differences and validation of the male typology. *Archives of General Psychiatry* 11:1248–1253.

Speer, G. S. (1940) The intelligence of foster children. *Journal of Genetic Psychology* 57:49–55.

von Knorring, A. L., Bohman, M. & Sigvardsson, S. (1982) Early life experiences and psychiatric disorders: An adoptee study. *Acta Psychiatrica Scandinavica* 65:283–291.

Chapter 4
Linkage and Association Studies

MICHAEL O'DONOVAN
PETER MCGUFFIN

Family, twin, and adoption studies that demonstrate the contribution of genetic factors toward the etiologies of the most common functional psychoses have been discussed in the preceding chapters—and provide perhaps the most definitive contribution to our understanding of causation. Although such studies are important in themselves, they necessarily lead on to further investigation with the ultimate aim of understanding the molecular basis of these disorders. In pursuit of this goal, two different types of marker studies have been performed; association studies and linkage studies. In this chapter we will outline the theoretical basis of such studies, the types of markers available for use, and the results of investigations to date. We will also consider some of the main obstructions that have hindered progress.

Linkage

The linkage strategy focuses on the cosegregation of a marker and a disease locus and attempts to detect deviation from independent assortment. Assuming that all positions on a chromosome are equally likely to be the subject of a recombination event, the closer two loci are the less likely they will become separated by recombination, and therefore the greater the departure from the 50% probability of them being inherited together. The statistical basis of linkage analysis is dealt with extensively in the chapter in this volume by Risch.

It is important to note that when two loci are said to be linked, this merely describes their relative proximity on a chromosome. It does not in any way imply a causal relationship between the marker locus and the disorder to which it is linked. In addition, one should not expect one particular form of the marker to be more commonly represented in those people with the disorder in comparison to the healthy population, just as after a pack of cards has been frequently shuffled, the queen of hearts is no more likely to be beside the jack of hearts than it is the jack of any other suit. However, exceptions to this expectation do occur and the phenomenon is called linkage disequilibrium. This most commonly arises when the disease and marker loci are very close, so that recombination events between the two are infrequent and only limited shuffling ensues. Linkage disequilibrium may also be the result of a relatively recent entry of one variant of the marker or the disorder into the gene pool, either through recent mutation or through mass migration.

Association

In association studies investigators measure the frequency of a marker phenotype in the diseased population and in the general population or other relevant control groups. Finding the phenotype marker to be more common (or less common) in those affected with the disorder may suggest that the marker phenotype is the result of a mutation which also contributes to the disorder (or protects against it). An alternative explanation is that the marker gene is extremely close to the disease gene and that the two are in linkage disequilibrium. However, a potential source of serious error is that of population stratification. If a disorder is particularly common in one subpopulation, an ethnic group, for example, and this is not allowed for in selecting the control sample, then all phenotypes common to that subpopulation will be more prevalent in the disease sample and, therefore, "associated." Thus if the detection of differences in marker frequencies unrelated to the disease is to be avoided, both the disease and control groups have to be very carefully matched. On the other hand, even if the marker and disease loci are in very close proximity, association may not be present if there is a high rate of mutation at either locus, or if they are in a chromosomal region that is particularly prone to recombination events (recombination hot spots). The issue of heterogeneity will be discussed later but this may bedevil this type of study, as well as linkage research.

Genetic Markers

Genetic markers are simple Mendelian traits that are *polymorphic*. That is, there are two or more alleles (or alternative genes) at the marker locus with a frequency of at least 1% in the population. It therefore follows that genetic markers are reliably detected and are not state dependent. In this strict sense genetic markers include many classical polymorphisms such as blood groups and human leukocyte antigen (HLA) types as well as DNA polymorphisms, but the term does not apply to other less clear-cut traits such as abnormalities of smooth-pursuit eye tracking or neuroendocrine status. Since the only matings that are informative for linkage are those in which one or both parents are heterozygous at both the marker and the disease loci (see the chapter by Risch), the degree to which a trait is polymorphic has a direct bearing on how useful it is in linkage research. The most useful genetic markers are those with a high polymorphism information content (pic), that is, those where an individual chosen at random has a high chance of being heterozygous.

It is worth noting that these desirable properties of simple inheritance (reliable detection, etc.) should not only apply to the marker(s) used to investigate linkage, but also to the disorder being studied itself. In reality we have no stable "trait" phenotype for either of the functional psychoses, other than "ever affected," and this is not always reliably discernible. At least as important, it has not been proven that any form of functional psychosis is a condition in which a single locus exerts a strong effect, although investigations into the mode of inheritance have not excluded this. It is clear, therefore, that schizophrenia and affective disorder, whether defined clinically or operationally, are less than ideal substrates for genetic marker study.

Genetic markers, then, can roughly be divided into "classical" genetic markers (e.g., ABO blood type rhesus status, HLA type, chromosome banding, protein polymorphisms, etc.) or DNA markers by which facets of genetic variability are examined more directly (see below).

DNA Markers

It is apparent from the nature of classical genetic markers that directly or indirectly they are products of structural genes (i.e., genes that encode polypeptides). Polymorphisms of the markers therefore reflect polymorphisms of the base pair coding sequences of these structural genes. However, since such genes account only for a very small proportion of the total genome, about 10%, only a fraction of the possible genetic diversity is tapped by studies of structural genes. The DNA code is said to be degenerate, by which it is meant that all of the amino acids (with the exception of tryptophan and methionine) are represented by more than one codon (the coding triplet of base pairs), and therefore many base pair mutations may have no functional effect. Those mutations which do have a functional effect may be disadvantageous, or even fatal, resulting in their removal from the genetic pool by evolutionary pressure. For these reasons, sequences of DNA that have no functional importance are likely to be richer sources of variation (Wolfe et al., 1989), and therefore if the polymorphisms in these regions could be detected they would be useful as genetic markers. As these mutations do not result in changes in protein products, they can only be detected by examining the DNA itself and comparatively recent advances in DNA technology have made this possible.

Restriction endonucleases (Meselson & Yuan, 1968) are bacterial enzymes with the ability to cleave double-stranded DNA (Smith & Wilcox, 1970) at sequence specific sites usually consisting of about six to eight base pairs. Taking Eco R1, for example, which recognizes the sequence GAATTC, wherever this occurs, the genome will be cut producing fragments, the length of which will be determined by the distances between these recognition sites. Mutations changing the DNA sequence may remove or create these sites. The result is that the number and length of fragments produced by the action of Eco R1 on a stretch of DNA will be variable within the population, that is, polymorphisms will exist. Such polymorphisms are called restriction fragment length polymorphisms (RFLPs) and most commonly occur because of single base pair mutations (Cooper & Schmidtka, 1984) but may also be produced by mutations that delete or insert large stretches of DNA.

After restriction enzyme digestion, fragments of DNA are then ordered according to size by electrophoresis on an agarose gel. The DNA, visualized at this stage with ultraviolet light, is seen as a smear, representing the millions of fragments of slightly different length produced by most endonucleases. At this point individual fragments cannot be recognized and therefore a means is required of highlighting pieces that originate from the marker locus.

First, the DNA is transferred and then bound to a nylon or nitrocellulose filter, the relative position of each fragment remaining unchanged from that in the gel (Southern blotting; Southern, 1975). After denaturing the DNA, the filter is exposed to single-stranded DNA probes complementary to the marker locus sequence to which, therefore, they will bind. If, as is usual, the probe is radiolabeled, the position of the polymorphisms can be determined by placing the filter paper next to X-ray paper (autoradiography), although a variety of techniques using non-radioactive labels are now available. From its position on the X-ray film, the size of the specific marker restriction fragments can be deduced, and as explained above, interindividual variability will be evident by variability in fragment size and hence position on the autoradiograph.

There are other recently developed techniques for detecting DNA polymorphisms, many of which promise to be even more efficient (Boerwinkle et al., 1989; Myers et al., 1985; Sheffield et al., 1989). However, only one other technique has as yet found a place in psychiatric genetics, a technique that has also had some impact in the popular press. This method is alternatively called minisatellite mapping, variable number tandem repeat mapping (VNTR), or DNA fingerprinting (Jeffreys et al., 1985). Some sequences of DNA have been discovered which consist of cores of 10–15 base pairs repeated in tandem, and these occur at multiple sites in the genome—these sequences are called hypervariable, or minisatellites (Jeffreys et al., 1985). The number of cores in tandem at any given site varies greatly throughout the population, more so in fact than the polymorphism of standard RFLPs. This great degree of polymorphism makes them potentially very valuable genetic markers, and the fact that the same core sequences occur at multiple sites (in the case of the "ALU" sequence approximately 300,000 different places) means that a probe for this sequence will "light up" fragments representing multiple loci which can therefore be simultaneously studied for linkage (Jeffreys et al., 1986). The main disadvantage of using these highly polymorphic markers is that the large number of fragments makes analysis more complex and the detection of linkage does not immediately define a unique area for the disease allele. However, if a fragment is identified that is linked to the disease, then it can be isolated and the flanking single copy DNA (i.e., the unique sequence still attached to the satellite) can then be used as a probe and be mapped directly to a chromosomal position.

Pitfalls of Marker Studies

The problems associated with the application of these markers in psychiatry are less to do with the markers themselves than with the illnesses to be studied. The most important difficulties are case definition, possible genetic heterogeneity, and phenotypic heterogeneity. Genetic heterogeneity implies that the same phenotype (e.g., depression) may be caused by different genetic disorders. Phenotypic heterogeneity on the other hand suggests that the same genetic abnormality may be manifest in a variety of clinical pictures, such as depression, mania, and depressive personality disorder.

Genetic Heterogeneity

If several genetic forms of a disorder exist, then in the absence of a way of differentiating between them it becomes increasingly difficult to detect any one. For example, if mutation at locus A or locus B could cause affective disorder, each being equally common, then when studying a random group of pedigrees it might be expected that about half of them would provide evidence of linkage to markers around locus A and about half would exclude such linkage. If there are 20 such genes the situation becomes correspondingly more complex. To add to the difficulties, if there are many different genes that cause affective disorder, a few of which are very common, it is conceivable that even within the same pedigree two affected individuals may have different mutant genes. One individual may therefore provide evidence favoring linkage to a marker but the other evidence against.

Phenotypic Heterogeneity

Phenotypic heterogeneity gives rise to problems largely because individuals are likely to be classified as unaffected even when they are abnormal. For example, if depressive personality disorder is truly a manifestation of the same mutation as mania, as long as only persons with a history of mania are accepted as affected, persons with depressive spectrum disorder who also have the marker polymorphism that suggests they should be affected will be falsely classified as recombinants, making it less likely that linkage will be detected.

To overcome these problems, we therefore need a reliable way of grouping cases in genetically homogeneous groups and of selecting a phenotype that is the best reflection of the underlying genotype. Unfortunately, we have no means of doing so on either clinical or laboratory grounds (Farmer & McGuffin, 1989; Kendell, 1987). At the moment, therefore, the problem of genetic heterogeneity, which we will come back to below, is unresolved, and our only solution to the problem of phenotypic heterogeneity is to include a variety of different case definitions in the analysis (Spence, 1987).

Other Problems

Another factor that complicates linkage analysis is incomplete penetrance (i.e., the probability of manifesting the disorder in those carrying a mutant gene is less than 100%). This results in some individuals being falsely ascribed as obligatory recombinants, again reducing the likelihood of linkage being detected.

Assortative mating, meaning that abnormal individuals are more likely to mate with phenotypically similar people, is likely to result in an increased likelihood of genetic heterogeneity for a disorder in a single pedigree. A birth cohort affect (Klerman et al., 1985), that is, that the onset of a disorder in recent generations may be earlier than in their ancestors, may result in difficulties compensating for the wide age range at which an apparently unaffected individual is still likely to have his first onset illness. Thus it is hazardous to ascribe a probability to individuals of being truly unaffected (or not yet affected).

For these reasons, the power of studies to detect linkage in psychiatric disorders may be quite low, and therefore, when faced with a study demonstrating linkage and another that fails to do so, or refutes it, it may be that an inherent bias toward false negative results is the explanation. On the other hand, there is also a risk of obtaining false positives because of multiple testing. This occurs because multiple markers are often used (Wiener, 1962), and the data are then analyzed assuming multiple different models of transmission and using multiple alternative definitions of disease phenotypes (McGuffin et al., 1990). These problems will be repeatedly encountered in the studies reviewed below and the solutions are unclear. It would, however, seem prudent to await independent replication before claims of linkage are viewed as anything other than preliminary. It has also been proposed that for psychiatric disorders the criteria lod score at which a finding is viewed to be significant be raised to 6 rather than 3 (Robertson, 1989), as is conventional in linkage studies of simple traits (Morton, 1955).

Genetic Marker Studies of Schizophrenia

Classical Marker Studies

Classical marker studies have focused primarily on blood grouping and the human leukocyte antigen (HLA) system though other polymorphic proteins have also been tested. We will look first at association and second at linkage studies.

There have been many association studies between schizophrenia and ABO blood groups (Mourant et al., 1975), although a major drawback is the absence of standardized diagnostic criteria. Among recent studies using strict diagnostic criteria no significant association between schizophrenia and any blood type has been found (McGuffin & Sturt, 1986; Rinieris et al., 1982).

The HLA system has provided a very fertile territory in detecting associations with common diseases that are somewhat analogous to the major psychoses, in that their etiology is poorly understood, their classification difficult, and their modes of inheritance uncertain. Encouraged by such findings, numerous studies on HLA and schizophrenia have been undertaken, which have been summarized by McGuffin and Sturt (1986). The only replicated associations were those with HLA-A9, B5, and the negative association with BW35. However, by subdividing schizophrenia into clinical subtypes, a consistent pattern has been detected for paranoid schizophrenia where out of nine studies, seven suggested HLA9 subtype to be more common in patients compared with controls. On pooling the data from the different centers, this difference was highly statistically significant (p = .003). It is notable that p values were obtained after a probably overstringent correction for the number of antigens tested (McGuffin & Sturt, 1986). The same reviewers also pooled the data on the three studies that suggested HLA1 to be more common in hebephrenic schizophrenia, but on this occasion statistical significance did not survive correction for multiple tests.

Another marker that has been closely examined for association with schizophrenia is the group specific component (Gc). Studies from Germany (Lange, 1982) found there was an association between Gc1-1 and schizophrenia, although other findings have not been consistent with this marker (Rudduck et al., 1985). Supportive independent replication is required before this association can be viewed with any confidence.

What, then, is to be made of the association between HLA9, which is by far the most statistically significant classical marker association, and schizophrenia? The reason behind such association with diseases is in general poorly understood, and although some elaborate theories have been proposed it would appear that while significant, the association is weak, contributing to perhaps about 1% of the variation in liability to paranoid schizophrenia (McGuffin & Sturt, 1986). The absence of a logical method of progressing from a finding of association or linkage between a disorder and a classical marker is a considerable drawback in their use, a disadvantage that is not shared with DNA markers as will be discussed later.

There have been fewer linkage studies with classical markers, probably a reflection of an increased degree of difficulty in both design and analysis. One pedigree has been reported in which albinism (unspecified type) appears to cosegregate with schizophrenia (Baron, 1976). More recently, a small family have been described (Clarke & Buckley, 1989) where out of three siblings, two have both albinism and a schizophrenic-like illness, the other individual being unaffected by either disorder. In themselves these are of limited interest but may provide direction for further investigation.

An early study (Turner, 1979) of linkage using HLA markers examined 13 families multiply affected by the broadly defined phenotype "schizotaxia" and found suggestive but not conclusive evidence of linkage to the HLA complex with a maximum lod score of 2.57 at a recombination fraction of .15. Since then there have been four studies (Andrew et al., 1987; Chadda et al., 1986; Goldin et al., 1987; McGuffin et al., 1983), all failing to replicate this finding and indeed excluding linkage up to a recombination fraction of .25 when the studies were combined (McGuffin, 1989). The reasons for the differences between first and subsequent studies is not clear. Heterogeneity could potentially be offered as an explanation, although on formal testing there is no evidence of its presence (McGuffin, 1989).

Three of the studies mentioned above (Andrew et al., 1987; McGuffin et al., 1983; Turner, 1979) also tested a range of other classical markers, none of which were found to be linked. It is worth noting here that for the Gc marker linkage was rejected (lod score greater than −2) up to a recombination fraction of .05. This would suggest that the positive association results of Lange (1982) cannot be explained on the basis of linkage disequilibrium.

Summarizing the findings for classical markers, it would appear that we are no closer to finding a locus or loci for schizophrenia. On the positive side, there may be a weak, but significant, association with HLA-A9, whereas the negative findings of linkage could allow about 6% of the genome to be excluded as a site of a dominant "schizophrenia" gene. The methodological difficulties described above, especially those relating to unknown mode of transmission, however, suggest that even this modest exclusion is only tentative.

DNA Markers

We have already seen how such polymorphic markers are generated. Before their application in schizophrenia is discussed, however, it would be useful to describe how they are used in general, and how, for the purpose of discovering the pathogenesis of a disorder, they are potentially of much greater value than the classical markers. Broadly speaking, there are two approaches to the use of DNA markers, a "random" search and a focus on "candidate" genes.

The random approach consists of using whatever markers happen to be available in the hope of detecting linkage with genes conferring disease susceptibility. Although this could be criticized as a crude "shot-gun" strategy, it has often yielded successes and the example most relevant to psychiatry was the finding of a linkage marker for Huntington's disease on the short arm of chromosome 4 (Gusella et al., 1983). The success in this case arose out of a study using just eight different "random" RFLPs but followed from a series of negative results from random linkage studies with many classical markers. A logical extension of the random approach, which has a greater probability of detecting major genes, but which is at the same time more expensive and time consuming, is to perform a systematic search. This systematic technique depends upon the creation of a genetic linkage map covering all of the human genome, such that every area of every chromosome will be linked to known RFLP marker loci (Botstein et al., 1980); the number of such loci required is estimated to be between 150 and 300. For any given disorder in which a major gene effect is supposedly important, markers will be selected from this map; by working through the whole series of markers, linkage can be pinpointed to one or more loci or excluded completely. As we have seen, however, such exclusion mapping may prove unreliable. This procedure is still termed "random" since no hypothesis about any particular chromosomal site or specific gene is required. Such a map covering 95% of the genome has been reported (Donis-Keller et al., 1987), although the number of RFLPs since then has expanded enormously. Thus this approach is practical, although it requires a great deal of energy, time, and finance.

The alternative approach, which has been widely applied in schizophrenia, is that of best guess or candidate markers. These are suitable not only for linkage studies but also for use in the association strategy. Candidate markers arise as a result of a suspicion that a particular chromosome, or even a particular gene itself, may contain a mutation of major influence in the genesis of the disorder. Attention to a particular chromosome may be suggested by finding chromosomal abnormalities cosegregating or associated with the disorder to be studied. The most obvious such association in psychiatry is that of trisomy 21 (Down's syndrome), which is in turn associated with Alzheimer's disease (Oliver & Holland, 1986). It was therefore logical to use markers already mapped to chromosome 21 as the starting point in investigating Alzheimer's disease, and indeed such an approach has already demonstrated linkage between chromosome 21 markers and Alzheimer's disease (Goate et al., 1989; St. George-Hyslop et al., 1987).

Alternatively, candidate genes may be suggested by current hypotheses regarding the etiology of a disease. Thus in psychiatric illness, the genes encoding enzymes important in the synthesis of neurotransmitters or their receptors are suitable candidates for investigation. DNA representing these genes could be used as probes in restriction fragment analysis instead of using randomly selected markers. The number of neuroreceptor clones that have been identified is rapidly increasing and includes cDNA for certain types of cholinergic, gabaergic, adrenergic, dopaminergic, serotonergic, as well as a variety of neuropeptide, receptors. Other candidates include genes involved in the control of the rate of translation of the aforementioned genes; glucocorticoid receptor genes, estrogen receptor genes, thyroid receptor genes, for example, and clones, are available for many of these.

A difficulty with the use of candidate genes is that our understanding of the pathophysiology of the psychoses is rudimentary at best. Given that this is so, it is likely that many of the candidates selected would be irrelevant and that our ignorance would lead us to overlook some potentially important genes. For this reason, rather than relying on an a priori understanding of the psychiatric disorders it seems likely that progress will be made in the opposite direction, a process called "reverse genetics" (McKusick, 1988). Random linkage is found first, followed by gene localization, sequencing, and identification of the abnormal gene products. Inferences can then be drawn about etiology. This approach has been spectacularly successful as applied to cystic fibrosis (Rommens et al., 1989).

Linkage Studies Using DNA Markers in Schizophrenia

Investigation of loci important in the etiology of schizophrenia was given what appeared to be a fortunate advance with the observations of chromosome 5 abnormalities in a Canadian family. A chance remark by a patient's mother led a team (Bassett et al., 1988) to investigate a family of Asian origin, two members of which (the proband and his maternal uncle) were noted to have both DSM-III schizophrenia and dysmorphic features. Chromosomal analysis revealed a balanced $^1/_5$ translocation (balanced meaning no excess portion of chromosome 5) in the proband's mother with the schizophrenic individuals both having unbalanced translocations resulting in a partial trisomy of chromosome 5. No other members of the family had either chromosome abnormalities or schizophrenia. The suggestion that chromosome 5 may be associated with schizophrenia was strengthened, albeit weakly, since a potential candidate gene, the glucocorticoid hormone receptor (GLR), had also been mapped to the same region of chromosome 5.

Following this lead, one team (Sherrington et al., 1988) studied five Icelandic and two British families with multiple affected members. A lod score of 6.49 for a dominant-like gene with high penetrance was obtained using multipoint analysis, the locus being between markers p105-599Ha and p105-153Ra. It also appeared that the inheritance of a gene in the 5q region could result not only in the variety of classical schizophrenia subtypes but also in a very wide "spectrum" of disorder. When a number of other psychiatric disorders, not generally thought to be related conditions (e.g., major depressive disorder), were included as "affected," there was an increase in the lod score. This finding is unexpected in that greatly broadening the definition of "affected" should increase the chance of including more false "cases" in the analysis and reduce the evidence for linkage.

Simultaneously, another study (Kennedy et al., 1988) reported on three related families from Northern Sweden that showed no evidence of linkage to markers in the 5q region, and when higher values of penetrance were assumed, the lod scores appeared to exclude linkage to this area. This Swedish sample, however, is a relatively genetically isolated community. In addition, the pedigree contained no individuals who met criteria for schizotypal or schizoid personality disorder. It can be argued, therefore, that the two groups are genetically dissimilar and may have different forms of the disorder (i.e., genetic heterogeneity). Since then, three other teams have reported, from Scotland (St. Clair et al., 1989), the United States (Detera-Wadleigh et al., 1989), and Wales (McGuffin et al., 1990); all reports were negative.

Controversy surrounding the interpretation of these reports has ensued. It has been suggested that the Scottish study was defective because some of the pedigrees contained bipolar affective illness, which was included in the affected phenotype (Gurling, 1990). In the most recent study, McGuffin et al. have combined all the published data, corrected inconsistencies in marker genotypes, and included the data from a missing pedigree from the work of Sherrington et al. Linkage is, then, resoundingly excluded from the combined data. Nonetheless, we are faced with the possibility that the one positive study is correct but has been swamped by the other negative studies. The possibility of true genetic heterogeneity appears to be slight, however, since positive linkage was obtained in the two British pedigrees as well as the Icelandic ones and therefore one would expect at least a minority of the pedigrees from the other studies to be transmitting such a widespread gene (McGuffin et al., 1990). If there is no such heterogeneity, then the four studies rejecting linkage can either be viewed as false negatives or the British-Icelandic study viewed as a false positive. As mentioned above, multiple testing of markers and inheritance models may increase the probability of a false positive. In this regard it is notable that the study of Sherrington et al. used many different schizophrenia phenotypes in addition to a variety of models of transmission. This increases the probability of a false positive, but there is, of course, no way of demonstrating this unequivocally. Therefore, it would again seem that the reasonable approach is the conservative one of assuming that any linkage is purely provisional until independent replication has been reported.

Marker Studies in Affective Disorder

The difficulties of defining the phenotype are particularly problematic with regard to the affective disorders. This is discussed in the chapter by Winokur and is the subject of a recent review (Farmer & McGuffin, 1989). For the purpose of our discussion it should be noted that we are considering only the more severe end of the affective disorder spectrum (if it is truly a spectrum).

Classical Markers

As for schizophrenia, most association studies have been concerned with blood groups and the HLA system. Some replicated reports indicate that blood group O is increased in bipolar patients compared with controls (Balgir, 1986; Masters, 1967; Mendlewicz et al., 1974; Parker et al., 1961; Rinieris et al., 1979). In other studies, group O has also been found to be increased in subtypes of unipolar depression (Irvine et al., 1965; Rinieris et al., 1979), as well as decreased (Balgir, 1986). Yet again group A has been associated with bipolar (Flemenbaum & Larson, 1976) and unipolar disorder (Balgir, 1986), whereas Beckman et al. (1978) found an association with group B and unipolar disorder. The lack of consistency in these studies (e.g., unipolar disorder being associated with blood groups A, B, and O) has led reviewers to reject any overall association (Goldin & Gershon, 1983) and to suggest that the conflicting results are due to population stratification (Flemenbaum & Larson, 1976).

The HLA system in association studies has provided no clearer a picture (Goldin & Gershon, 1983; McGuffin & Sargeant, 1991), with both sets of reviewers finding no convincing HLA association. Much of the apparent contradiction in association studies of this type again relates to the statistical problems of multiple markers and population stratification (Targum et al., 1979).

Linkage studies using classical markers have provided more interesting results, perhaps pointing toward linkage to the X chromosome. In the interests of continuity, discussion of these will be deferred until the section on DNA markers. Despite a provocative report of possible HLA linkage (Weitkamp et al., 1981) we can now conclude that certain autosomal markers, such as the HLA system, ABO grouping, haptoglobin group specific component, and glutamine pyruvate transaminase, have been effectively excluded from distances of at least 10 centimorgans (Goldin et al., 1983; Johnson et al., 1981; Kidd et al., 1984; Waters et al., 1988). However, the MNS blood group locus gives a modest positive score from three of the studies and may be worth further pursuit (McGuffin & Sargeant, 1991).

DNA Markers and Affective Disorder

The first apparent breakthrough using DNA markers for either of the functional psychoses was made in a study of a large pedigree of the Old Order Amish (OOA) community in Pennsylvania (Egeland et al., 1987). This religious sect was, for many reasons, ideal for study of psychiatric disorder since all of its 12,000 members were documented to have descended from only 30 founder individuals. One might expect, therefore, if ever one was to achieve genetic homogeneity in a sample, that this would be it. In addition, strict social pressure against the use of drugs and alcohol reduces the likelihood of including "substance induced" phenocopies, while large family size and geographic concentration of related individuals improve pedigree ascertainment.

Initially, negative results were reported on 41 RFLP markers (Kidd et al., 1987) but, finally, concentrating on one large pedigree, evidence was found of linkage between a gene for affective disorder, the Harvey-ras-1 (Hras) oncogene locus, and the insulin (INS) locus in the p15 region of chromosome 11. The maximum lod score for this linkage was 4.9.

The excitement of this finding, and its possible generalization to other populations, was immediately tempered by the simultaneous publication of two other negative studies (Detera-Wadleigh, Berettini et al., 1987; Hodgkinson et al., 1987) reporting lod scores of less than −2 (i.e., excluding linkage) in Icelandic and North American Irish pedigrees respectively. In addition, Hodgkinson's team used a DNA probe for the tyrosine hydroxylase gene (also mapped to 11p), this too being excluded as a candidate in their pedigrees. In attempting to explain these findings, the application of formal testing for heterogeneity was performed (McGuffin, 1988) and the results strongly supported the heterogeneity hypothesis. This seemed conceivable given the singular nature of the sample that had generated the positive lod score. However, the study of the OOA was not static, and with the addition of follow-up material, the positive linkage finding began to disintegrate (Kelsoe et al., 1989). Two individuals who had been unaffected became ill, immediately becoming obligatory recombinants. Furthermore, the pedigree was extended with the addition of another familial branch in which linkage was excluded to either Harvey-ras or insulin loci. When combined with the original pedigree and with the diagnostic changes, linkage could be excluded up to a distance of 15 centimorgans. It has been suggested that this updated negative finding may be due to the inclusion of a branch that is carrying a different gene for affective disorder. While this is possible, it seems more reasonable to conclude that there is no gene for affective illness in the chromosome 11p15 region.

X-Linkage

Since publication of the first study by Reich et al. (1969), there have been repeated suggestions that X-linked genes may be involved in the transmission of affective disorder, with the marker trait being color blindness. A further study (Mendlewicz & Fleiss, 1974) supported this finding and also detected linkage with the Xg blood group locus. Subsequently, linkage between another

X-chromosome marker, which is close to the color blindness locus, glucose 6 phosphate dehydrogenase deficiency, and affective disorder was reported (Mendlewicz et al., 1980), apparently enhancing the case for X-linked inheritance. However, it later emerged that the Xg locus is too far distant from the color blindness locus for an affective illness gene to be linked to both. Furthermore, a large United States-European collaborative study (Gershon et al., 1980) failed to find overall evidence of linkage between bipolar illness and color blindness even though the data from one center were strongly positive. Whether this was due to true heterogeneity or the result of a systematic error in sampling cases was the subject of much debate. Two recent reports, however, have added strength to the claims of X-linkage. Baron et al. (1987) confirmed linkage with a maximum lod score of 7.52–7.97, depending upon the model assumed. Baron's study of Israeli families was closely followed by a report of linkage (lod 3.10) in Belgian families (Mendlewicz et al., 1987) between RFLPs for blood clotting factor 1X (F9) and affective disorder. A peculiar feature of these families is that they were chosen because of absence of father-son transmission, but in fact in only two cases does an affected male have any children at all.

Although these findings have certainly supported the X-linkage hypothesis of affective disorder, they can only account for a minority of cases (Bucher et al., 1981). As is the general pattern, studies (quoted by Mandel et al., 1989) are emerging, finding no evidence of X-linkage. Another problem (recapitulating the Xg-color blindness story) is that the F9 and GGPD sites are so far apart that it is not likely that one disease is linked to both of them (Mandel et al., 1989). Thus one would have to postulate that there are at least two genes for affective disorder on the X chromosome (and three if the possibility is still allowed of Xg linkage). While this may be the case, if such a high degree of heterogeneity exists, then it does not bode well for marker studies in affective disorder.

Candidate Gene Studies in Schizophrenia and Affective Disorder

Candidate gene studies are of obvious appeal but depend for their feasibility first upon the cloning of relevant genes and second on the probability that the current ideas concerning the biochemical basis of the functional psychoses are correct. It might not be surprising, therefore, if positive results were not obtained or were slow in coming. Several such studies have been undertaken, but at the time of writing most of these are as yet unpublished in complete form.

The first candidate gene study to be reported was an association study between RFLPs for pro-opiomelanocortin gene (POMC) and both schizophrenia and bipolar disorder (Feder et al., 1985). The POMC gene encodes the precursor peptide of ACTH, and beta endorphin, both of which are potential candidates for the psychoses. However, no association between POMC RFLPs in either schizophrenia or affective disorder (32 and 22 cases, respectively) was found. The interpretation is limited by the small numbers, leaving it still possible that the gene plays a role in either or both of the disorders, but in pedigrees not sampled by this study. Further problems are that any association will weaken if mutation rates of the disease are high, or if population frequencies of marker alleles are not known with confidence. Another study (Detera-Wadleigh, de Miguel et al., 1987) tested RFLPs of the neuropeptide Y and somatostatin genes for linkage to affective disorder in two informative pedigrees with a total of 15 affected members. The results were not encouraging but were only exclusive for linkage if a rare dominant transmission with complete penetrance model was assumed. Also, RFLPs for gastrin-releasing peptide, substance P, adenosine de-aminase, and neuropeptide Y were tested for association with schizophrenia and affective disorder. No differences in genotypic frequencies were found, but if the patient group was combined there was an association with adenosine de-aminase that reached borderline significance. As has been evident from the work with classical markers, results that are of borderline significance merit cautious further study rather than excitement.

Other candidate genes that have been excluded by linkage studies include a number of adrenergic receptor subtypes (Hoehe et al., 1989) and tyrosine hydroxylase (Hodgkinson et al., 1987) in affective disorder; linkage between schizophrenia and the dopamine D_2 receptor (Moises et al., 1989), tyrosine hydroxylase, and homeobox -2 genes (Kennedy et al., 1989) has similarly been excluded. When considering the implications of these studies it is important to remember that the exclusion of a candidate in a linkage study only applies to that pedigree or set of pedigrees, not to the population of affected people as a whole.

Conclusion

It will be evident to the reader that the present state of genetic marker research in schizophrenia and affective disorder is confused. On the one hand, several positive results have emerged for both major categories of psychosis, but on the other, there have been a disappointing failure of replication and various disparities in reported findings, which do not encourage strong confidence in any of the linkages suggested to date. There is, however, cause for optimism. The mapping of the human genome is proceeding rapidly and there are already sufficient polymorphisms available to mount a systematic search throughout most chromosomes for major genes for the major psychoses. This will present a major effort and if properly coordinated must involve multiple centers working in collaboration. Formal collaboration programs have already been set up in Europe under the auspices of the European Science Foundation and in the United States under the National Institute of Mental Health. In addition, informal collaborative ventures are under way in North America and elsewhere. All of this effectively takes psychiatry into an era of potential "big science." As we have suggested earlier, the search for major genes for psychosis is not premised on any strong evidence that they exist, but rather, on the fact that there is no compelling evidence against their existence. That is, there is almost certainly an important genetic contribution to both schizophrenia and manic depression but both might still be polygenic syndromes. If, however, major genes do exist they are certain to be eventually located by linkage strategies. Therefore, if detection of well-established and well-replicated linkages is going to occur it will occur within the next five to ten years.

References

Andrew, B., Watt, D. C., Gillespie, C. & Chapel, H. (1987) A study of genetic linkage in schizophrenia. *Psychological Medicine* 17:363–370.

Balgir, R. S. (1986) Serological markers in unipolar and bipolar affective disorders. *Human Heredity* 36:250–253.

Baron, M. (1976) Albinism and schizophreniform psychosis: A pedigree study. *American Journal of Psychiatry* 133:1070–1073.

Baron, M., Risch, N., Hamburger, R., Mandel, B., Kushner, S., Newman, M., et al. (1987) Genetic linkage between X-chromosome markers and bipolar affective illness. *Nature* 326:289–292.

Bassett, A. S., McGillivray, B. C., Jones, B. D. & Pantzar, J. T. (1988) Partial trisomy chromosome 5 cosegregating with schizophrenia. *Lancet* i:799–801.

Beckman, L., Cedergren, B., Perris, C. & Strandman, E. (1978) Blood groups and affective disorders. *Human Heredity* 28:48–55.

Boerwinkle, E., Xiong, W., Fourest, E. & Chan, L. (1989) Rapid typing of tandemly hypervariable loci by the polymerase chain reaction: Application to the apolipoprotein B 3' hypervariable region. *Proceedings of the National Academy of Science* 86:212–216.

Botstein, D., White, R. L., Skolnick, M. & Davis, R. W. (1980) Construction of a genetic linkage map in man using restriction fragment length polymorphisms. *American Journal of Human Genetics* 32:314–331.

Bucher, K. D., Elston, R. C., Green, R., Whybrow, P., Helzer, J., Reich, T., et al. (1981) The transmission of manic depressive illness: II. Segregation analysis of three sets of family data. *Journal of Psychiatric Research*, 16:65–78.

Chadda, R., Kullhara, P., Singh, T. & Sehgal, P. S. (1986) HLA antigens in schizophrenia: A family study. *British Journal of Psychiatry* 149:612–615.

Clarke, D. J. & Buckley, M. E. (1989) Familial association of albinism and schizophrenia. *British Journal of Psychiatry* 155:551–553.

Cooper, D. N. & Schmidtka, J. (1984) DNA restriction fragment length polymorphisms and heterozygosity in the human genome. *Human Genetics* 66:1–16.

Detera-Wadleigh, S. D., Berettini, W. H., Goldin, L. R., Boorman, D., Anderson, S. & Gershon, E. S. (1987) Close linkage of c-Harvey-ras-1 and the insulin gene to affective disorder is ruled out in three North American pedigrees. *Nature* 325:806–808.

Detera-Wadleigh, S. D., de Miguel, C., Berrettini, W. H., DeLisi, L. E., Goldin, L. R. & Gershon, E. S. (1987) Neuropeptide gene polymorphisms in affective disorder and schizophrenia. *Journal of Psychiatric Research* 21:581–587.

Detera-Wadleigh, S. D., Goldin, L. R., Sherrington, R., Encio, I., de Miguel, C., Berrettini, W., et al. (1989) Exclusion of linkage to 5q11-13 in families with schizophrenia and other psychiatric disorders. *Nature* 339:391–393.

Donis-Keller, H., Green, P., Helms, C., Cartinhour, S., Weiffenbach, B., Stephens, K., et al. (1987) A genetic linkage map of the human genome. *Cell* 51:319–337.

Egeland, J. A., Gerhard, D. S., Pauls, D. L., Sussex, J. N., Kidd, K. K., Allen, C. R., et al. (1987) Bipolar affective disorders linked to DNA markers on chromosome 11. *Nature* 325:783–787.

Farmer, A. E. & McGuffin, P. (1989) The classification of the depressions: Contemporary confusion revisited. *British Journal of Psychiatry* 155:437–443.

Feder, J., Gurling, H. M. D., Darby, J. & Cavalli-Sforza, L. L. (1985) DNA restriction fragment analysis of the proopiomelanocortin gene in schizophrenia and bipolar disorders. *American Journal of Human Genetics* 37:286–294.

Flemenbaum, A. & Larson, J. W. (1976) ABO-Rh blood groups and psychiatric diagnosis: A critical review. *Diseases of the Nervous System* 37:581–583.

Gershon, E. S., Mendlewicz, J., Gastpar, M., Bech, P., Goldin, L. R., Kielholz, P., et al. (1980) A collaborative study of genetic linkage of bipolar manic depressive illness and red/green colorblindness. *Acta Psychiatrica Scandinavica* 61:319–338.

Goate, A. M., Owen, M. J., James, L. A., Mullan, M. J., Rossor, M. N., Haynes, A. R., et al. (1989) Predisposing locus for Alzheimer's disease on chromosome 21. *Lancet* i:352–355.

Goldin, L. R., De Lisi, L. F. & Gershon, E. S. (1987) The relationship of HLA to schizophrenia in 10 nuclear families. *Psychiatry Research* 20:69–78.

Goldin, L. R. & Gershon, E. S. (1983) Association and linkage studies of genetic marker loci in major psychiatric disorders. *Psychiatric Developments* 4:387–418.

Goldin, L. R., Gershon, E. S., Targum, S. D., Sparkes, R. S. & McGinniss, M. (1983) Segregation and linkage analyses in families of patients with bipolar, unipolar and schizo-affective mood disorders. *American Journal of Human Genetics* 35:274–288.

Gurling, H. M. D. (1990) Review. *Transmission* 9:15.

Gusella, J. F., Wexler, N. S., Coneally, P. M., Naylor, S. L., Anderson, M. A., Tanzi, R. E., et al. (1983) A polymorphic DNA marker genetically linked to Huntington's disease. *Nature* 306:234–238.

Hodgkinson, S., Sherrington, R., Gurling, H., Marchbanks, R., Reeders, S., Mallet, J., et al. (1987) Molecular genetic evidence for heterogeneity in manic depression. *Nature* 325:805–806.

Hoehe, M. R., Berrettini, W. H., Leppert, M., Lalouel, J. M., Goldin, L. R., Byerley, W. (1989) Adrenergic receptor genes: Polymorphisms, mapping, linkage studies in manic-depressive pedigrees. Presented at 1st World Congress of Psychiatric Genetics, Cambridge.

Irvine, D. G. & Miyashita, H. (1965) Blood groups in relation to depressions and schizophrenia. *Canadian Medical Association Journal* 92:551.

Jeffreys, A. J., Wilson, V. & Thein, S. L. (1985) Hypervariable "minisatellite" regions in human DNA. *Nature* 314:67–73.

Jeffreys, A. J., Wilson, V., Thein, S. L., Weatherall, D. & Ponder, A. J. (1986) DNA "fingerprints" and segregation analysis of multiple markers in human pedigrees. *American Journal of Human Genetics* 39:11–24.

Johnson, G. F. S., Hunt, G. E., Robertson, S. & Doran, T. J. (1981) A linkage study of manic depressive disorders with HLA antigens, blood groups, serum proteins and red cell enzymes. *Journal of Affective Disorders* 3:43–58.

Kelsoe, J. R., Ginns, E. I., Egeland, J. A., Gerhard, D. S., Goldstein, A. M., Bale, S. J., et al. (1989) Re-evaluation of the linkage relationship between chromosome 11p loci and the gene for bipolar affective disorder in the Old Order Amish. *Nature* 342:238–243.

Kendell, R. E. (1987) Diagnosis and classification of functional psychoses. *British Medical Bulletin* 43:499–513.

Kennedy, J. L., Giuffra, L. A., Moises, H. W., Cavalli-Sforza, L. L., Pakstis, A. J., Kidd, J. R., et al. (1988) Evidence against linkage of schizophrenia to markers on chromosome 5 in a Northern Swedish pedigree. *Nature* 336:167–170.

Kennedy, J. L., Giuffra, L. A., Wetterberg, L., Sjogren, L., Cavalli-Sforza, L. L., Pakstis, A. J. (1989) Schizophrenia in Swedish kindred is not linked to D5S6 Tyrosine Hydroxylase or the homeobox −2 locus. Presented at 1st World Congress of Psychiatric Genetics, Cambridge.

Kidd, J. R., Egeland, J. A., Pakstis, A. J., Castiglione, C. M., Pletcher, B. A., Morton, L. A. & Kidd, K. K. (1987) Searching for a major genetic locus for affective disorder in the Old Order Amish. *Journal of Psychiatric Research* 21(4):577–580.

Kidd, K. K., Egeland, J. A., Molthan, L., Pauls, D. L., Kruger, S. D. & Messner, K. H. (1984) Amish Study, IV: Genetic linkage study of pedigrees of bipolar probands. *American Journal of Psychiatry* 141:1042–1048.

Klerman, G. L., Lavori, P. W., Rice, J., Reich, T., Endicott, J., Andreasen, N. C., et al. (1985) Birth cohort trends in rates of major depressive disorder among relatives of patients with affective disorder. *Archives of General Psychiatry* 42:689–693.

Lange, V. (1982) Genetic markers for schizophrenia sub-groups. *Psychiatria Clinica* 15:133–144.

Mandel, J.-L., Willard, H. F., Nussbaum, R. L., Romeo, G., Puck, J. M. & Davies, K. E. (1989) Report of the committee on the genetic constitution of the X chromosome. *Cytogenetic Cell Genetics* 51:384–437.

Masters, A. B. (1967) The distribution of blood groups in psychiatric illness. *British Journal of Psychiatry* 113:1309–1315.

McGuffin, P. (1988) Major genes for major affective disorder? *British Journal of Psychiatry* 153:591–596.

McGuffin, P. (1989) Genetic markers: An overview and future perspectives. In *A Genetic Perspective for Schizophrenic and Related Disorders*, ed. E. Smeraldi & L. Belloni. Milan: Edi-Ermes.

McGuffin, P., Festenstein, H. & Murray, R. M. (1983) A family study of HLA antigens and other genetic markers in schizophrenia. *Psychological Medicine* 13:31–43.

McGuffin, P. & Sargeant, M. (1991) Genetic markers and affective disorder. In *New Genetics of Mental Illness*, ed. P. McGuffin & R. M. Murray, pp. 163–179. London: Heinemann.

McGuffin, P., Sargeant, M., Hett, G., Tidmarsh, S., Whatley, S. & Marchbanks, R. M. (1990) Exclusion of a schizophrenia susceptibility gene from the chromosome 5q11-q13 region. New data and a reanalysis of previous reports. *American Journal of Human Genetics* 47:524–535.

McGuffin, P. & Sturt, E. (1986) Genetic markers in schizophrenia. *Human Heredity* 36:65–88.

McKusick, V. A. (1988) The morbid anatomy of the human genome. *Medicine* 67:1–19.

Mendlewicz, J. & Fleiss, J. L. (1974) Linkage studies with X chromosome markers in bipolar (manic depressive) and unipolar (depressive) illness. *Biological Psychiatry* 9:261–294.

Mendlewicz, J., Linkowski, P. & Wilmotte, J. (1980) Linkage between glucose-6-phosphate dehydrogenase deficiency and manic-depressive psychosis. *British Journal of Psychiatry* 137:337–342.

Mendlewicz, J., Massart-Guiot, T., Wilmotte, J. & Fleiss, J. L. (1974) Blood groups in manic depressive illness and schizophrenia. *Diseases of the Nervous System* 35:39–41.

Mendlewicz, J., Simon, P., Sevy, S., Charon, F., Brocas, H., Legros, S. & Vassart, G. (1987) Polymorphic DNA marker on X chromosome and manic depression. *Lancet*, May 30:1230–1232.

Meselson, M. R. & Yuan, M. (1968) DNA restriction enzyme from *E. coli*. *Nature* 217:1110–1114.

Moises, H. W., Gelertner, J., Grandy, D. K., Giuffra, L. A., Kidd, J. R., Pakstis, A. J., et al. (1989) Exclusion of the D-2 dopamine receptor gene as candidate gene for schizophrenia in a large pedigree from Sweden. Presented at 1st World Congress on Psychiatric Genetics, Cambridge.

Morton, N. E. (1955) Sequential tests for the detection of linkage. *American Journal of Human Genetics* 7:277–318.

Mourant, A. E., Kopec, A. C. & Domaniewska-Sobczak, K. (1975) *Blood Groups and Diseases*. Oxford: Oxford University Press.

Myers, R. M., Lumelsky, N., Lerman, L. S. & Maniatis, T. (1985) Detection of single base substitutions in total genomic DNA. *Nature* 313:495–497.

Oliver, C. & Holland, A. J. (1986) Down's syndrome and Alzheimer's disease: A review. *Psychological Medicine* 16:307–322.

Parker, J., Theile, A. & Spielberger, C. (1961) Frequency of blood types in a homogenous group of manic depressive patients. *Journal of Mental Sciences* 107:936–942.

Reich, T., Clayton, P. J. & Winokur, G. (1969) Family history studies v. the genetics of mania. *American Journal of Psychiatry* 125:1358–1369.

Rinieris, P. M., Stefanis, C. N., Lykouras, E. P. & Varsou, E. K. (1979) Affective disorders and ABO blood types. *Acta Psychiatrica Scandinavica* 60:272–278.

Rinieris, P., Stefanis, C., Lykouras, E. & Varsou, E. (1982) Subtypes of schizophrenia and ABO blood types. *Neuropsychobiology* 8:57–59.

Robertson, M. (1989) False start on manic depression. *Nature* 342:1989.

Rommens, J. M., Iannuzzi, M. C., Kerem, B-S., Drumm, M. L., Melmer,

70 M. O'DONOVAN, P. McGUFFIN

G., Dean, M., et al. (1989) Indentification of the cystic fibrosis gene: Chromosome walking and jumping. *Science* 245:1058–1059.

Rudduck, C., Franzen, G., Hansson, A. & Rorsman, B. (1985) Gc serum groups in schizophrenia. *Human Heredity* 35:11–14.

Sheffield, V. C., Cox, D. R., Lerman, L. S. & Myers, R. M. (1989) Attachment of a 40-base-pair G+C-rich sequence (GC-clamp) to genomic DNA fragments by the polymerase chain reaction results in improved detection of single-base changes. *Proceedings of the National Academy of Science* 86:232–236.

Sherrington, R., Brynjolfsson, J., Petursson, H., Potter, M., Dudleston, K., Barraclough, B., et al. (1988) Localization of a susceptibility locus for schizophrenia on chromosome 5. *Nature* 336:164–167.

Smith, H. O. & Wilcox, K. W. (1970) A restriction enzyme from haemophilus. Influenzae: I. Purification and general properties. *Journal of Molecular Biology* 51:379–391.

Southern, E. M. (1975) Detection of specific sequences among DNA fragments separated by gel electrophoresis. *Journal of Molecular Biology* 98:503–517.

Spence, M. A. (1987) Genetic linkage: Sampling issues and multipoint mapping. *Journal of Psychiatric Research* 21(4):631–637.

St. Clair, D., Blackwood, D., Muir, W., Baillie, D., Hubbard, A., Wright, A. & Evans, H. J. (1989) No linkage of 5q11-q13 markers to schizophrenia in Scottish families. *Nature* 339:305–309.

St. George-Hyslop, P. H., Tanzi, R. E., Polinski, R. J., Haines, J. L., Nee, L., Watkins, P. C., et al. (1987) The genetic defect causing familial Alzheimer's disease maps on chromosome 21. *Science* 235:885–890.

Targum, S. D., Gershon, E. S., Van Eerdewegh, M. & Rosentine, N. (1979) Human leukocyte antigen system not closely linked to or associated with bipolar manic depressive illness. *Biological Psychiatry* 14(4):615–636.

Turner, W. J. (1979) Genetic markers for schizotaxia. *Biological Psychiatry* 14:177–205.

Waters, B., Sengar, D., Marchenko, I., Rock, G., Lapierre, Y., Forster-Gibson, C. J. & Simpson, N. E. (1988) A linkage study of primary affective disorder. *British Journal of Psychiatry* 152:560–562.

Weitkamp, L. R., Stancer, H. C., Persand, E., Flood, C. & Guttormsen, S. (1981) Depressive disorders and HLA: A gene on chromosome 6 than can affect behaviour. *New England Journal of Medicine* 305:1301–1306.

Wiener, A. S. (1962) Blood groups and disease. *Lancet* i:813–816.

Wolfe, K. H., Sharp, P. M. & Li, W. H. (1989) Mutation rates differ among regions of the mammalian genome. *Nature* 337:283–285.

Chapter 5

Genetic Linkage
Studies in Psychiatry:
Theoretical Aspects

NEIL RISCH

The past decade has witnessed a surge of interest in human genetic diseases, primarily due to remarkable advances in molecular genetic technology. Ten years ago, one would have been considered lucky to identify the chromosomal location of a genetic disease locus. Even then, the practical implications of such a finding, such as genetic counseling and prenatal diagnosis, were limited by the paucity of polymorphic genetic markers. Today, however, the human genome is rapidly being covered by a very large number of highly polymorphic marker loci, known as restriction fragment length polymorphisms (RFLPs) (Botstein et al., 1980). These polymorphisms are identified through variation in DNA sequence among individuals. Bacterial restriction enzymes, which cleave double-stranded DNA only when a specific DNA sequence appears, are used to digest an individual's DNA. Sequence variation among individuals means that the pattern of lengths of DNA strands remaining after digestion will differ. The cleaved DNA is hybridized to a previously identified unique strand of DNA called a probe; the probe usually as a known chromosomal location. The size of the fragment(s) that hybridize to the probe can be identified. If there is sequence variation in the region of the probe such that some individuals have a cleavage recognition site for the restriction enzyme while others do not, distinct patterns of fragment lengths among individuals will appear. It is the presence or absence of these cleavage sites that determine the genetic polymorphism. Another type of DNA polymorphism, referred to as variable number tandem repeat (VNTR), has also been shown to be extremely useful for genetic linkage studies in humans (Nakamura et al., 1987). In this case, the locus of interest is characterized by the tandem repetition of a constant length DNA sequence. Differences among individuals

71

are due to variation in the number of times the constant sequence is tandemly repeated. The distribution of alleles in the population for VNTR loci often contains a very large number of fragment sizes, sometimes approaching a continuous distribution. Hence, some of these loci are the most polymorphic known.

The advent of RFLPs and VNTRs in the early and mid 1980s allowed the chromosomal localization of many genetic disease loci, such as Huntington's disease (Gusella et al., 1983), neurofibromatosis (Barker et al., 1987; Rouleau et al., 1987), polycystic kidney disease (Reeders et al., 1985), familial adenomatous polyposis (Bodmer et al., 1987; Leppert et al., 1988), and cystic fibrosis (Beaudet et al., 1986), to name but a few. The large number of such marker loci and their highly polymorphic nature provided genetic counseling and prenatal diagnosis with a precision previously unknown. However, the most remarkable and exciting advances took place toward the end of the decade, with the demonstration that it was possible to move from the chromosomal localization of a disease gene to the actual identification of the faulty gene itself and the characterization of its product. With the identified protein product in hand, it is possible to understand disease etiology. The first successful examples of this approach occurred for the X-linked recessive disease Duchenne (and Becker) muscular dystrophy (Burghes et al., 1987; Koenig et al., 1987; Monaco et al., 1986) and the autosomal dominantly inherited childhood eye tumor retinoblastoma (Friend et al., 1986; Fung et al., 1987; Lee et al., 1987). Both of these diseases, however, had the advantage that a proportion of cases were associated with observable chromosomal deletions, which were used to rapidly localize the faulty gene. Such an approach would not apply to the vast majority of genetic diseases, which are not associated with observable chromosomal aberrations. Therefore, perhaps the most important announcement of the past decade was the identification of the gene for cystic fibrosis (CF), the commonest recessive disease in Caucasians (Kerem et al., 1989; Riordan et al., 1989; Rommens et al., 1989). Because no observable chromosomal abnormalities had been identified with CF, sophisticated molecular genetic techniques, such as chromosome jumping and walking had to be employed. The most important implication of this finding, aside from the obvious relevance for understanding the etiology of CF, was that it demonstrated that genetic linkage analysis could lead to the identification of the faulty gene for a typical genetic disease using currently available technology, and therefore that other genetic diseases, in theory, could be analyzed in a similar fashion.

Although the 1980s were a decade of excitement and progress in human genetics in general, with remarkable success of the genetic linkage approach to so many diseases, it was perhaps a decade of disappointment in psychiatric genetics. Several claims of genetic linkage of major psychiatric disorders to particular chromosomal regions failed to be replicated. The disappointment may have been due to a failure to recognize a critical fact: the diseases that were successfully mapped were among the simplest to analyze genetically, whereas psychiatric illnesses are among the most complicated. In this chapter, I describe theoretical and statistical aspects of genetic linkage analysis, with particular reference to its application to genetically complex diseases such as psychiatric illnesses.

The Phenomenon of Linkage

The phenomenon of linkage derives from the fact that genes coexist on chromosomes. In humans there are 22 pairs of non-sex chromosomes (called autosomes) and 1 pair of sex chromosomes, which are 2 X chromosomes in females and an X and Y chromosome in males. In meiosis, the 2 chromosomes in a pair segregate, so that only 1 chromosome of each pair is transmitted to a gamete. The various chromosomes segregate independently, so that alleles that are located on different chromosomes will also segregate independently. This means that two traits encoded by loci on different chromosomes will be inherited independently within families. By contrast, if two traits are encoded by alleles on the same chromosome (called syntenic), they may not be transmitted independently to gametes, but will tend to be transmitted together.

Theoretically, one might think that syntenic alleles would always be transmitted together because they are on the same chromosome. However, in meiotic disjunction, chromosomes do not stay intact; crossing over, or recombination, occurs between the two homologous chromosomes of a pair. When at least one crossover occurs between the two loci of interest, they will appear to segregate independently, as if they were on different chromosomes.

The frequency with which crossovers occur between two loci depends on their distance apart. For loci that are close, crossing over may be rare; for loci far apart on a chromosome, crossing over may be a certainty. Therefore, if loci are sufficiently far apart on a chromosome, they will segregate independently, as if they were on different chromosomes. If the loci are nearby, they will not segregate independently, but also, because of recombination, they will not always segregate together, unless they are extremely close. The recombination fraction θ denotes the proportion of all gametes that are recombined between the two loci of interest.

Statistical Basis For Linkage Detection

The statistical basis for the detection of linkage is grounded in simply inherited, Mendelian traits, that is, those whose genetic variation can be attributed to segregation at a single locus. The simplest traits are those termed codominant, for which there is a one-to-one correspondence between genotype and phenotype. For example, consider a locus A with two alleles A_1 and A_2. The locus is codominant provided the three genotypes A_1A_1, A_1A_2 and A_2A_2 are phenotypically distinguishable. Suppose we have a second codominant locus B with alleles B_1 and B_2. The phenomenon of linkage is observed when alleles at two loci do not assort independently at meiosis.

Meiotic events in an individual are inferred by the observed genotypes in his/her offspring. The specific allele transmitted to each child will be determinable only if the parent is heterozygous (e.g., A_1A_2); if the parent is homozygous (e.g., A_1A_1), all offspring will receive the same allele (A_1). Therefore, in order to track the segregation of alleles at two loci, an informative parent must be heterozygous at both loci (e.g., $A_1A_2B_1B_2$). Even if one parent is doubly heterozygous, all offspring may not be fully informative if the other parent is heterozygous for the same alleles. For example, if both parents are A_1A_2 and a child is A_1A_2, the A_1 allele (and similarly the A_2 allele) is equally likely to have been transmitted by either parent.

An additional complication occurs when we do not know a priori which alleles in a doubly heterozygous parent occur together

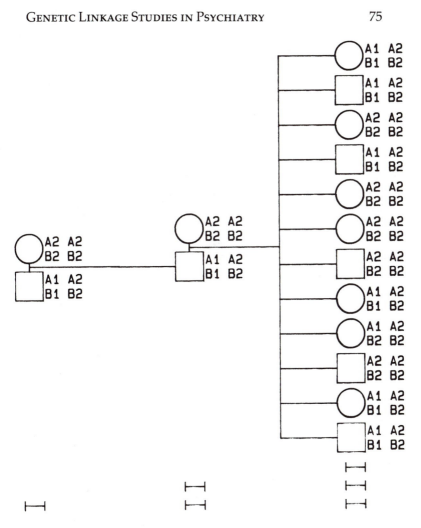

FIGURE 5.1.
Example of a pedigree with genetic marker loci for a linkage study

on the same chromosome (referred to as "phase"). For example, if an individual is $A_1A_2B_1B_2$, the chromosomal constitution may be A_1B_1 and A_2B_2, or conversely A_1B_2 and A_2B_1. These difficulties are circumvented in experimental organisms because parents and matings can be created with any desired chromosomal configuration. For example, consider the pedigree given in Figure 5.1. The father II-1 was created from a mating of a double heterozygote with an $A_2A_2B_2B_2$ double homozygote. Therefore, we know his phase; his maternal chromosome contains A_2B_2, while his paternal chromosome contains A_1B_1. Because the mother II-2 is doubly homozygous $A_2A_2B_2B_2$, the 12 offspring in generation III will be fully informative for meiotic events that occurred in the father, and completely uninformative for such events in the mother. Each offspring has inherited a chromosome containing A_2B_2 from the mother II-2, so the two alleles inherited from the father can always be deduced. In 11 of the 12 offspring, an unrecombined chromosome (either A_1B_1 or A_2B_2) has been inherited from the father. Offspring III-4, however, has inherited a recombined chromosome A_1B_2 from her father. Therefore, among the 12 offspring, 1 is recombinant and 11 parental (nonrecombinant). By examining enough such matings and offspring, one can obtain a direct count of the proportion of offspring that are recombinant. In the absence of linkage, the expectation is that half will be recombinant. Therefore, if significantly fewer than half the offspring are recombinant, evidence for linkage is obtained.

Linkage analysis in humans is inherently more complicated because of the inability to artificially create desired matings. One must sample existing matings and families until informative cases are obtained. At a minimum, for a family to be informative for linkage, one parent must be doubly heterozygous for the loci of interest. Even if such a parent is identified, his/her phase may be unknown; however, if there are at least two offspring, the family may still be informative. As opposed to experimental organisms, in humans it is usually not possible to directly count recombinant and nonrecombinant offspring. Therefore, a more general linkage assessment approach, applicable to any informative family, is required. A solution to this problem was given by Morton (1955). For each family, a statistic called a "lod score" can be derived. Lod scores are summable across families, and hence the evidence for linkage can be obtained from the total lod score summed across all families.

The basis for the lod score is as follows. For a given family, the likelihood (or probability) of obtaining the observed segregation pattern of the two traits can be derived in terms of the recombination fraction θ; refer to this likelihood as $L(\theta)$. A similar likelihood can be obtained under the assumption of no linkage; this is equivalent to assuming $\theta = 1/2$, so the likelihood assuming no linkage is $L(1/2)$. A likelihood ratio $LR(\theta)$, or odds ratio, can be obtained by dividing $L(\theta)$ by $L(1/2)$. Now suppose that we have collected n different informative families, and that the likelihood of the i-th such family is $L_i(\theta)$. Because the n families are independent, the likelihood (probability) of the entire collection of n families is merely $L_1(\theta)L_2(\theta) \ldots L_n(\theta)$, the product of the individual likelihoods. Similarly, the likelihood ratio for the entire collection of families is the product of the individual likelihood ratios, namely $LR(\theta) = (L_1(\theta)/L_1(1/2))(L_2(\theta)/L_2(1/2)) \ldots (L_n(\theta)/L_n(1/2))$. It is often simpler mathematically to deal with sums than products, so the common log of $LR(\theta)$, or $\log_{10}LR(\theta)$ is taken. The lod score is then defined as the $\log_{10}LR(\theta)$, and this is why lod scores are summable across families (the log of a product is the sum of the logs of the individual terms).

As an example, suppose by chance we have obtained a human pedigree resembling the one shown in Figure 5.1. In this case, we have 12 offspring fully informative for meiotic events occurring in the father. Because 11 of the 12 are nonrecombinant and 1 is recombinant, the probability of this family is proportional to $\theta(1 - \theta)^{11}$. Therefore, for this family $LR(\theta) = \theta(1 - \theta)^{11}/(1/2)\times(1/2)^{11} = 2^{12}\theta$ $(1 - \theta)^{11}$, and $\log_{10}LR(\theta) = 12\log2 + \log\theta + 11\log(1 - \theta)$. For example, at $\theta = 0$, $\log_{10}LR(\theta) = -\infty$; this value corresponds to the fact that one recombination has definitely occurred, and hence a value $\theta = 0$ is impossible. At $\theta = .1$, $\log_{10}LR(\theta) = 2.09$; at $\theta = .2$, $\log_{10}LR(\theta) = 1.83$.

As another example, reconsider the pedigree in Figure 5.1, but suppose that the grandparents in generation I were not available. In this case, the phase of the father II-1 is unknown, so the offspring cannot be definitively assigned as recombinants or nonrecombinants. It is usually assumed that the doubly heterozygous father is equally likely a priori to be in each of the two possible phases (A_1B_1 and A_2B_2 or A_1B_2 and A_2B_1). Therefore, among the offspring, there are either 11 nonrecombinants and 1 recombinant or 11 recombinants and 1 nonrecombinant, with each possibility having a prior probability of $1/2$. Therefore, the likelihood of such a family is $(1/2)[\theta(1 - \theta)^{11} + \theta^{11}(1 - \theta)]$, and the likelihood ratio $LR(\theta) = (1/2)[\theta(1 - \theta)^{11} + \theta^{11}(1 - \theta)]/(1/2)^{12} = 2^{11}[\theta(1 - \theta)^{11} + \theta^{11}(1 - \theta)]$. The lod score at $\theta = 0$, as before, is $-\infty$, because at least one definite recombination has occurred; at $\theta = .1$, the lod score is 1.79 and at $\theta = .2$, the lod score is 1.53.

A large value for the lod score $\log_{10}LR(\theta)$ for a given value of θ indicates that the observed data are much more likely to have occurred under the assumption of linkage with a recombination fraction of θ (<1/2) than under the assumption of no linkage (θ = 1/2). But how large is large enough? The answer to this question needs to be placed in the context of the prior probability that the two loci of interest are linked within a certain distance of one another. Elston and Lange (1975) showed that the prior probability that two randomly chosen autosomal genes lie on the same chromosome within a recombination fraction of 30% of each other is about 2%. Morton (1955) determined that an appropriate lod score criterion for concluding linkage between two loci would be one such that the posterior frequency of false positives (i.e., false linkage claims) should be 5%. Taking into account the low prior probability of linkage (about 2%), he showed theoretically (assuming families are analyzed sequentially) that a lod score of 3 is required to obtain a false positive rate of 5%. A lod score of 3 at a given value θ (<1/2) means that the observed data are 1,000 times more likely to have occurred if the loci are linked at the given value of θ than if the loci are unlinked (θ = 1/2). In a subsequent analysis, Rao et al. (1978) showed empirically that the lod score criterion of 3 was, in fact, effective at maintaining a 5% false positive rate. As a further protection against false claims of linkage, a linkage must be replicated in a second, independent laboratory before it is considered confirmed.

The example in Figure 5.1 depicts two genetic traits that are codominant. Often, one or both of the traits of interest has a more complicated pattern of inheritance, especially when one is a disease. Most genetic diseases are characterized as either dominant or recessive. For the dominant case, possession of a single copy of a mutant allele is sufficient for disease expression. Dominant diseases tend to be very rare, and nearly all affected individuals are heterozygous for the disease allele (homozygotes being extremely rare). If compatible with reproduction, a dominant disease may appear in many consecutive generations of a pedigree, where, on average, half the offspring of an affected individual will be affected. Such pedigrees can be extremely useful in linkage studies, because all affected individuals are heterozygous at the disease locus, and therefore their offspring will be informative for linkage provided the affected parent is also heterozygous at the marker locus (second trait) of interest. Spouses are usually well, and therefore homozygous for the normal allele at the disease locus. As an example, consider again the pedigree in

Figure 5.1. Suppose that the A locus codes for the disease, where A_1 is the disease allele and A_2 the normal allele. Then individuals I-1, II-1, III-1,2,4,5,9,11,12 with genotype A_1A_2 are affected, and the remainder, with genotype A_2A_2 are normal.

By contrast, for a recessive disease, homozygosity for the mutant allele is required for disease expression. Recessive diseases also tend to be rare, but in some cases can exist in the population at moderately high frequency, because the majority of the mutant alleles in a population are carried by heterozygotes, who are unaffected by the disease, and, in some cases, may in fact have a selective advantage over the normal homozygotes. Whereas selection acts very effectively and rapidly in reducing the frequency of a deleterious dominant allele, selection against a recessive allele is much slower. For a recessive disease, the most common family constellation is two normal parents (who are both heterozygotes) and one or several affected offspring. In such families, the risk to offspring is 25%. Also, there is an increased probability of parental consanguinity, especially if the disease is rare. In a typical family, both parents will be (normal) heterozygotes, and therefore theoretically informative for linkage provided they are also heterozygous at the marker of interest. In this case, however, the unaffected offsprings' genotypes at the disease locus will be equivocal because they can be either a heterozygote or normal homozygote, each of which corresponds to a normal phenotype. By contrast, affected offspring are disease allele homozygotes, and therefore are known to have received the disease allele from each heterozygous parent. Hence, affected offspring are usually much more informative for linkage than their unaffected sibs. An additional problem is that phase in the parents is generally unknown, because all grandparents are normal and there is no way to identify which grandparent transmitted the disease allele. However, when there is parental consanguinity and grandparents are available for marker determination, it is possible to derive phase because the grandparent transmitting the disease alleles can be identified. In general, for a recessive disease, pedigrees involving consanguinity can be among the most useful for linkage studies.

For both the dominant and recessive cases, the relationship between genotype and phenotype is known, although it is not one-to-one. For the dominant case, individuals heterozygous or homozygous for the disease allele are affected, while normal homozygotes are unaffected. For the recessive case, only disease allele homozygotes are affected, while heterozygotes and normal homozygotes are unaffected. For some genetic diseases, individuals who possess a disease-producing genotype remain unaffected, or the age at which an individual first shows symptoms may vary greatly. These phenomena are known as reduced and age-dependent penetrance, respectively. Provided the penetrance (probability of being affected) of a disease genotype is known (including age-dependent penetrances), such complications pose no difficulty for linkage analysis by the lod score approach. In this case, as previously, the genotypes of affected individuals will be known with certainty; it is for unaffected individuals that there is increased ambiguity about genotype. However, using a designated (possibly age-dependent) penetrance function, one can assign a probability of each possible disease locus genotype for every unaffected individual in a pedigree, and use these values when calculating lod scores.

   In conclusion, it is important to recognize that the lod score method of linkage analysis and the significance criterion of 3.0 are based on a number of assumptions: that the two traits of interest are inherited as simple, single locus Mendelian traits; that the relationship between genotype and phenotype is known (e.g., dominance and penetrance relationships); that gene frequencies of the various alleles are known. All these parameters must be specified to calculate lod scores and tests of linkage accurately.

Multipoint Analysis

Prior to the last decade, multipoint linkage analysis was a technique restricted largely to experimental species. With the advent of molecular genetic methodology for creating a large number of polymorphic markers saturating a specific chromosomal region, multipoint analysis in humans has taken on new relevance. The primary importance of multipoint analysis lies in (1) increased power to detect linkage of a disease locus, or conversely to exclude a disease locus from a specific chromosomal region; (2) the precise localization of a disease locus to a short stretch of DNA; and (3) increased precision in genetic counseling and prenatal diagnosis.

   When a disease is rare, the number of useful families available for linkage analysis may be small, so maximal advantage of the existing families must be taken. As discussed in a previous section, in order for an individual to be informative for linkage analysis, he/she must be heterozygous at both loci of interest. For a rare dominant disease, affected individuals are heterozygous at the disease locus; for a recessive disease, the unaffected parents of an affected individual are heterozygous. In these cases, the limitation will be the frequency of heterozygosity at the marker locus of interest. For example, if for a given marker only 50% of the population is heterozygous while the other 50% is homozygous, only half, on average, of the potentially useful individuals will actually be informative. The majority of genetic markers have heterozygote frequencies far below 100%. Therefore, one approach to increasing the informativeness of a sample is to use several different marker loci near one another in a specific chromosomal region. For example, if we have four loci, each with a heterozygote frequency of 50%, the probability that a given individual will be heterozygous for at least one such locus is $1 - (.5)^4 = 93.75\%$; hence, in this case, nearly all individuals heterozygous at the disease locus will also be heterozygous at a marker.

A problem now arises in that different informative individuals will be heterozygous for different marker loci; some might be heterozygous for two or three, while others might be heterozygous for one only. The question is how to evaluate linkage evidence from such a collection of families and markers. Whereas in two-point analysis the hypothesis being tested statistically is that the two loci are linked with a recombination fraction of θ versus the hypothesis of no linkage ($\theta = 1/2$), in multipoint analysis the hypothesis is that the disease locus lies somewhere in the vicinity of a set of linked marker loci versus entirely outside this region. It is generally assumed that the order and distance apart of the marker loci are known. The task is therefore to calculate the probability of an observed constellation of disease and marker phenotypes in a given family, assuming the disease locus is situated at a specific position with respect to the marker loci (the linkage hypothesis), as well as the same probability assuming that the disease locus is entirely unlinked to the group of marker loci. As in two-point analysis, the $\log_{10}$ of the ratio of these probabilities is the lod score, and the same significance level of 3.0 to conclude linkage is used. The calculation of multipoint linkage probabilities can be very complicated, and depend on mathematical models of the process of crossing over along a chromosomal segment. In a long enough interval, multiple crossovers can occur and need to be taken into account in analyses. Furthermore, locations of crossovers are not independent, adding an additional layer of complexity, although many analyses incorporate the simplifying assumption that multiple crossovers occur independently. Fortunately, efficient computer algorithms are generally available for performing these difficult calculations.

In two-point analysis, lod scores are calculated for several values of θ and a lod score curve can be drawn. In multipoint analysis, a similar curve can be drawn, where the horizontal axis is now the location of the disease locus with respect to the markers. An example is given in Figure 5.2. This example is from the analysis of simulated data created for Problem 2 of Genetic Analysis Workshop 6, as reported by Elston et al. (1989). The results are for a dominantly inherited disease locus located in a cluster of three marker loci. Figure 5.2a shows a two-point lod score curve for marker locus 3 with the disease. The maximum lod score is 45 and occurs at a θ value of .04. This result suggests that the disease

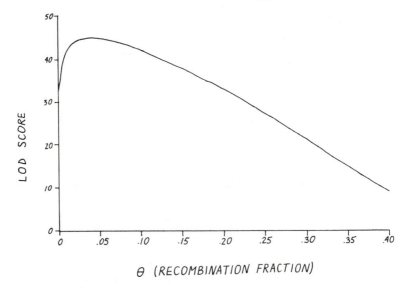

FIGURE 5.2a.
Pairwise lod score curve for the dominant disease locus with marker locus 3

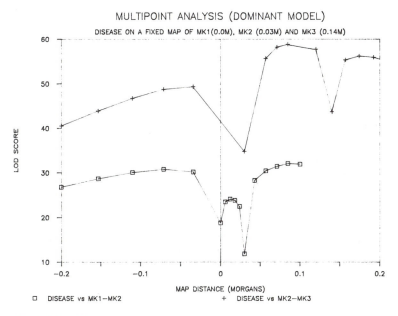

FIGURE 5.2b.
Multipoint lod score curve for the dominant disease locus with
three linked marker loci

locus is linked to marker 3, and the recombination fraction between loci is reasonably in the range of .01% to .09%. Figure 5.2b
shows the results of multipoint analysis using all three markers.
The known order of the three markers is 1-2-3, and the recombination fraction between markers 1 and 2 is .03 and between
markers 2 and 3 it is .10. The figure shows the lod score values
when the disease locus is situated at various positions. The maximum lod score of 58 occurs when the disease locus is between
marker loci 2 and 3. The disease locus is placed between markers 2
and 3 with some certainty because the lod score for the next most
likely location is 3 less (or 1,000 times less likely). Two important
points should be derived from Figure 5.2. First, the lod score for
linkage of the disease locus to this region of the chromosome is
greater in the multipoint analysis compared to the two-point
analyses; this is due to the fact that the number of individuals informative (heterozygous) for linkage is increased when multiple
markers are used. Second, the multipoint figure illustrates that,

in theory, a disease locus can be reliably placed in a short segment of a chromosome.

Genetic Heterogeneity

For Mendelian diseases, it is common that the specific mutant alleles causing disease in the population are not of unique origin. Often, distinct mutations occur at the same locus; however, sometimes mutations at different loci can lead to the same disease phenotype. The phenomenon of distinct mutant alleles at the same locus is referred to as allelic, or intralocus, heterogeneity, whereas that of mutant alleles at different loci is referred to as nonallelic, or interlocus, heterogeneity; both forms of variation are subsumed under the general notion of genetic heterogeneity—that is, that distinct mutant alleles can lead to the same disease.

Genetic heterogeneity can manifest in several ways. First, a single genetic disease entity can sometimes be subcategorized into clinically homogeneous subgroups that "breed true" within families; these subgroups are attributable to distinct mutations, either at the same locus or different loci. For example, the autosomal dominant disorder neurofibromatosis can be subdivided clinically into two major subgroups, the more common vonRecklinghausen type (NF1), and the less common acoustic type (NF2); both forms run true within families (Riccardi & Eichner, 1986). The fact that genetic (nonallelic) heterogeneity underlies this division was revealed by linkage studies that demonstrated that the locus causing NF1 is located on chromosome 17 (Barker et al., 1987) and the locus causing NF2 is located on chromosome 22 (Rouleau et al., 1987). An example of allelic heterogeneity is given by the X-linked recessive muscular dystrophies; the more severe form is referred to as the Duchenne type, and the milder form the Becker type; each type "runs true" within families. It is now known that both forms result from mutations at the same genetic locus, with disease severity a function of the severity of the mutation (Kunkel et al., 1986).

A second way in which genetic heterogeneity can be revealed is by subgrouping based on distinct Mendelian patterns of inheritance. For Example, the genetic eye disease retinitis pigmentosa can manifest in families in an autosomal dominant, autosomal recessive, or X-linked recessive fashion (Boughman et al., 1980). Certainly, the X-linked form involves a locus (or loci) that is distinct from the autosomal forms; the autosomal dominant and recessive forms very likely involve distinct loci as well.

A third way in which genetic heterogeneity can be revealed is through linkage studies. A single disease entity with a uniform pattern of inheritance may still be caused by mutant alleles at different loci. Upon performing linkage studies, some families may give strong evidence for linkage to a particular chromosomal region, while other families given clear evidence against linkage to the same region. The classic example of this type of evidence is the blood disease elliptocytosis, which is uniformly inherited in an autosomal dominant fashion. Linkage studies with the blood group marker Rh revealed some families closely linked to the marker, whereas the remainder were clearly unlinked, presumably due to a mutation at a different locus (Morton, 1956). The linked and unlinked forms are now recognized as differing biochemically as well (Palek, 1985).

In conclusion, genetic heterogeneity is the phenomenon that distinct mutant alleles can lead to the same disease phenotype. The different mutations may occur at the same or different loci, may have different patterns of inheritance, and may or may not be associated with distinct clinical and/or biochemical pictures. It is a concept that applies specifically to Mendelian diseases.

Linkage Analysis of Genetically Complex Diseases

All of the foregoing discussion involved linkage analysis of genetically simple diseases, that is, those whose inheritance pattern can be explained by genetic variation at a single locus. Such diseases are generally rare, a result of selection acting to keep deleterious allele frequencies low. However, many common, chronic diseases are also familial, and are believed to have a genetic component. Examples include common cancers, such as those of the breast and colon, diabetes, coronary heart disease, epilepsy, and major psychiatric disorders, to mention just a few. By and large, these disorders do not follow simple patterns of Mendelian inheritance, and are generally much more common than Mendelian diseases. For these cases, environmental influences play a greater role (for example, as evidenced by low monozygotic twin concordance). The genetic component may be complex, involving many loci that contribute to disease risk, and the effect of any single locus on modifying risk is not that large. Although the genetic basis for Mendelian (single locus) diseases is well understood, for the common complex diseases it is not. It is therefore important to assess the role of genetic linkage analysis for such disorders.

The major issue when studying the genetics of complex diseases is the number of loci that may be involved, the magnitude of their effects, and possible interactions among the effects of different loci. One measure of the total magnitude of genetic effects is the ratio of risk to relatives compared to population prevalence (Risch, 1987). For Mendelian diseases this ratio (denoted λ_R) is generally very high (perhaps 100 or greater), because of the high risk to relatives and low population prevalence. By contrast, for the common complex diseases, λ_R values are much smaller; for example, for the major psychiatric disorders such as schizophrenia and bipolar illness, λ_R for first-degree relatives is about 10. The pattern of λ_R values across different types of relative can also offer possible clues about number of loci and magnitude of gene effects. Risch (1990a) has shown that for single-locus traits, $\lambda_R - 1$ should decrease by a factor of 2 with each degree of genetic relationship—for example, from first- to second- to third-degree relatives. A decrease by a factor greater than 2 indicates the presence of multiple loci with interaction effects among loci, and therefore rules out both single locus models of inheritance, as well as simple models of genetic heterogeneity (for example, the retinitis pigmentosa model described above) because genetic heterogeneity excludes interaction effects among loci. When λ_R for first-degree relatives is not too small, for example greater than 5, some discrimination among models is possible. For example, Risch (1990a) showed that family recurrence data for schizophrenia appear to be more compatible with a few loci with moderate effects and interactions among loci than either a single locus model or a genetic heterogeneity model.

Although overall recurrence risks to relatives is one useful approach, it does not reveal the entire picture. For example, a modest risk to relatives might be attributable to two very different models: (1) several or many loci, with common allele frequencies, no one of which has a major impact on modifying disease risk; or (2) one locus with a rare genotype conferring high risk, mixed in with a preponderance of nongenetic cases (or "phenocopies"). Complex segregation analysis, in which one fits genetic models to family data, is an important technique that can theoretically distinguish between these possibilities, and is often (and should be) performed before undertaking linkage studies. Which model is more tenable can have an important impact on how to perform linkage studies. For example, if model (2) were correct, one would attempt to identify exclusively those families segregating the high risk allele for linkage studies; these are likely to be the ones with the greatest preponderance of ill members. By contrast, if model (1) were correct, it might be preferable to study small constellations of affected relatives, such as pairs.

For psychiatric disorders, mode of inheritance studies, such as segregation analysis, have given little evidence for the existence of single major locus effects. A negative segregation analysis result, however, does not entirely rule out the possibility that such loci exist, although they are likely to be an uncommon cause of disease. Therefore, many investigators have proceeded in the hope of identifying such loci through linkage studies. This has primarily taken the form of identifying families with many individuals affected, under the assumption that these are the most likely to be segregating a highly penetrant dominant allele.

At this stage, the definition of the affected phenotype becomes a critical issue. In some cases, a subset of a complex disease can be defined on the basis of clinical, demographic, and/or biochemical criteria, where the subset is only a minor component of the total disease and is therefore rare in the general population. Further, this subset, narrowly defined, shows a Mendelian pattern of inheritance. One example is Alzheimer's disease (AD), which is very common by the eighth decade of life, but extremely rare in the fourth and fifth decades (Martin et al., 1988). Yet several very large, multigenerational pedigrees with 20–50 individuals affected with AD before age 60 and an autosomal dominant pattern of inheritance have been reported. For example, the pedigrees analyzed by St. George-Hyslop et al. (1987) were of this type. As another example, consider coronary heart disease (CHD). As in AD, CHD is extremely common at later ages, but relatively uncommon before age 40. CHD before age 40 is also much more familial than later onset disease (Risch et al., 1985; Rissanen,

1979). One subset of early onset CHD can be attributed to hyper-cholesterolemia, with individuals having total serum cholesterol concentrations in excess of 300 even at early ages. This combination of early onset CHD, hypercholesterolemia, and additional clinical features such as tendon xanthomas, defines a rare, Mendelian subgroup of CHD that is inherited in an autosomal dominant fashion (Goldstein et al., 1973). It is now known that this form is caused by mutations of the cholesterol receptor locus (Brown & Goldstein, 1974), and indeed one family has demonstrated linkage between hypercholesterolemia and the cholesterol receptor locus on chromosome 19p (Leppert et al., 1986). As another example, cleft lip is a common familial, non-Mendelian birth defect. However, several rare Mendelian syndromes including cleft lip as a component have been defined; one such syndrome, Van der Woude syndrome (known to be an autosomal dominant condition), has been mapped to chromosome 1q through linkage studies (Murray et al., 1988).

In each of these examples of common diseases in which linkage analysis has been successful in locating a disease locus, it has been for a rare subset of disease with a Mendelian pattern of inheritance well-defined prior to the linkage study. Although it may be possible to characterize the etiology of these conditions through linkage analysis and subsequently through reverse genetics, it cannot automatically be assumed that the remaining cases will have a similar etiology. For example, the etiology of early CHD due to mutations in cholesterol receptor genes is well-defined, but the remainder of CHD, and in particular early-onset CHD, is probably regulated largely through different mechanisms.

For psychiatric conditions, no well-defined Mendelian subsets exist. In fact, to obtain pedigrees with many affected individuals, the criteria for affected status are usually *broadened* instead of narrowed. For example, linkage studies of bipolar affective illness (manic-depression) usually include cases of unipolar illness (major depression), which may have a population lifetime prevalence as high as 10% (Robins et al., 1984). Similarly, a recent study of schizophrenia (Sherrington et al., 1988) included a spectrum of psychiatric conditions with a population frequency of greater than one-third. When the definition of affected needs to be broadened to include highly prevalent conditions in order to achieve large, multiplex pedigrees, the Mendelian nature of such pedigrees, although possible, becomes less secure. This fact, coupled with assortative mating for psychiatric illnesses (Merikangas, 1982) and recently observed secular trends in disease rates (Klerman et al., 1985), has made linkage studies in psychiatry difficult.

In the absence of a well-defined Mendelian form of a disease, the conventional approach to linkage analysis and its statistical interpretation no longer apply. Without a well-defined Mendelian pattern of inheritance, optimal sampling strategies for detecting disease susceptibility loci are not obvious. One approach is to focus on a large sample of affected relative pairs and consider sharing of marker alleles identical by descent by such pairs (Bishop & Williamson, 1990; Risch, 1990b). At present, if larger constellations of affected relatives, such as pedigrees, are sampled, the optimal method of analysis is unclear. One approach is to examine marker identity by state among affected pairs in the pedigree (Weeks & Lange, 1988), but this approach, while robust to mode of inheritance assumptions, loses much of the potential linkage information available in such pedigrees (Risch, 1990c; Risch et al., 1989). Alternatively, one can arbitrarily specify one or two modes of inheritance (such as dominant and recessive with reduced penetrance) and perform a conventional linkage analysis (Risch et al., 1989). This approach should not seriously increase the likelihood of producing a false positive result (Clerget-Darpoux et al., 1986), and would take account of more of the linkage information in the pedigree than other robust methods.

Furthermore, the lod score criterion of 3.0 and its corresponding 5% false positive rate are based on the prior assumption of a major, single locus effect. When no major locus effect exists, all positive linkage results will be false. Therefore, any lod score or other statistical finding must be interpreted in the context of one's prior belief in the existence of a single major locus effect.

For psychiatric diseases the absence of a prior Mendelian inheritance pattern causes linkage results to take on greater significance than they do for conventional Mendelian conditions, because they are interpreted to "prove" there is a genetic component to the disease, whereas Mendelian diseases require no such proof. However, if one is to conclude such "genetic proof" on the basis of linkage alone, it would seem that *stricter* statistical criteria than those applied to conventional Mendelian diseases are necessary, and consistent replication of a reported linkage finding is critical. Absence of replication should not be casually attributed to genetic heterogeneity.

In conclusion, psychiatric diseases have complex, mysterious etiologies. From an epidemiologic perspective, perhaps the strongest risk factor for developing such a condition is having a positive family history. By and large, family, twin and adoption studies have supported the hypothesis that genetic factors contribute to disease etiology. Although recent developments in molecular biology have enabled the possibility of understanding the disease process through reverse genetics, their application to complex disorders such as behavioral ones will be complicated and difficult, and substantial progress may be some time in coming.

References

Barker, D., Wright, E., Nguyen, K., Cannon, L., Fain, P., Goldgar, D., et al. (1987) The gene for von Recklinghausen neurofibromatosis is in the pericentromeric region of chromosome 17. *Science* 236:1100–1102.

Beaudet, A., Bowcock, A., Buchwald, M., Cavalli-Sforza, L. L., Farrall, M., King, M.-C., et al. (1986) Linkage of cystic fibrosis to two tightly linked DNA markers: Joint report from a collaborative study. *American Journal of Human Genetics* 39:681–693.

Bishop, D. T. & Williamson, J. A. (1990) The power of identity by state methods for linkage analysis. *American Journal of Human Genetics* 46:254–265.

Bodmer, W. F., Bailey, C., Bodmer, J., Bussey, H., Ellis, A., Gorman, P., et al. (1987) Localization of the gene for familial adenomatous polyposis. *Nature* 328:614–616.

Botstein, D., White, R. L., Skolnick, M. & Davis, R. W. (1980) Construction of a genetic linkage map in man using restriction fragment length polymorphisms. *American Journal of Human Genetics* 32:314–331.

Boughman, J. A., Conneally, P. M. & Nance, W. E. (1980) Population genetic studies of retinitis pigmentosa. *American Journal of Human Genetics* 32:223–235.

Brown, M. S. & Goldstein, J. L. (1974) Familial hypercholesterolemia: Defective binding of lipoproteins to cultured fibroblasts associated with impaired regulation of 3-hydroxy-3-methylglutaryl coenzyme A reductase activity. *Proceedings of the National Academy of Sciences* 71:788–792.

Burghes, A. H. M., Logan, C., Hu, V., Belfall, B., Worton, R. G. & Ray, P. N. (1987) A cDNA clone from the Duchenne/Becker muscular dystrophy gene. *Nature* 328:434–437.

Clerget-Darpoux, F., Bonaiti-Pellie, C. & Hochez, J. (1986) Effects of mis-specifying genetic parameters in lod score analysis. *Biometrics* 42:393–399.

Elston, R. C. & Lange, K. (1975) The prior probability of autosomal linkage. *Annals of Human Genetics* 38:341–350.

Elston, R. C., MacCluer, J. W., Hodge, S. E., Spence, M. A. & King, R. H. (1989) Genetic analysis workshop 6: Linkage analysis based on affected pedigree members. In *Multipoint Mapping and Linkage Based upon Affected Pedigree Members: Genetic Analysis Workshop 6*, ed. R. C. Elston, M. A. Spence, S. E. Hodge & J. W. MacCluer, pp. 93–103. New York: Alan R. Liss.

Friend, S. H., Bernards, R., Rogelj, S., Weinberg, R. A., Rapaport, J. M. & Dryja, T. P. (1986) A human DNA segment with properties of the gene that predisposes to retinoblastoma and osteosarcoma. *Nature* 323:643–646.

Fung, Y.-K., Murphree, A. L., T'Ang, A., Qian, J., Hinrichs, S. H. & Benedict, W. F. (1987) Structural evidence for the authenticity of the human retinoblastoma gene. *Science* 236:1657–1661.

Goldstein, J. L., Schrott, H. G., Hazzard, W. R., Bierman, E. L. & Motulsky, A. G. (1973) Hyperlipidemia in coronary heart disease. II. Genetic analysis of lipid levels in 176 families and delineation of a new inherited disorder, combined hyperlipidemia. *Journal of Clinical Investigation* 52:1544–1568.

Gusella, J. F., Wexler, N. S., Conneally, P. M., Naylor, S. L., Anderson, M. A., Tanzi, R. E., et al. (1983) A polymorphic DNA marker genetically linked to Huntington's Disease. *Nature* 306:234–238.

Kerem, B., Rommens, J. M., Buchanan, J. A., Markiewicz, D., Cox, T. K., Chakravarti, A., et al. (1989) Identification of the cystic fibrosis gene: Genetic analysis. *Science* 245:1073–1080.

Klerman, G. L., Lavori, P. W., Rice, J., Reich, T., Endicott, J., Andreasen, N. C., et al. (1985) Birth cohort trends in rates of major depressive disorder among relatives of patients with affective disorder. *Archives of General Psychiatry* 42:689–693.

Koenig, M., Hoffman, E. P., Bertelsen, C. J., Monaco, A. P., Feener, C. & Kunkel, L. M. (1987) Complete cloning of the Duchenne muscular dystrophy (DMD) cDNA and preliminary genomic organization of the DMD gene in normal and affected individuals. *Cell* 53:219–228.

Kunkel, L. M., Hejtmancik, J. F., Caskey, C. T., Speer, A., Monaco, A. P., Middlesworth, W., et al. (1986) Analysis of deletions in DNA from patients with Becker and Duchenne muscular dystrophy. *Nature* 322:73–77.

Lee, W.-H., Bookstein, R., Hong, F., Young, L.-H., Shew, J.-Y. & Lee, E.Y.-H.P. (1987) Human retinoblastoma susceptibility gene: Cloning, identification and sequence. *Science* 235:1394–1399.

Leppert, M., Dobbs, M., Scambler, P., O'Connell, P., Nakamura, Y.,

Stauffer, D., et al. (1988) The gene for familial polyposis coli maps to the long arm of chromosome 5. *Science* 238:1411–1413.

Leppert, M. F., Hasstedt, S. J., Holm, T., O'Connell, P., Wu, L., Ash, O., et al. (1986) A DNA probe for the LDL receptor gene is tightly linked to hypercholesterolemia in a pedigree with early coronary disease. *American Journal of Human Genetics* 39:300–306.

Martin, R. L., Gerteis, G. & Gabrielli, W. F. (1988) A family-genetic study of dementia of Alzheimer type. *Archives of General Psychiatry* 45:894–900.

Merikangas, K. R. (1982) Assortative mating for psychiatric disorders and psychological traits. *Archives of General Psychiatry* 39:1173–1180.

Monaco, A. P., Neve, R. L., Colletti-Feener, C., Bertelsen, C. J., Kurnit, D. M. & Kunkel, L. M. (1986) Isolation of candidate cDNAs for portions of the Duchenne muscular dystrophy gene. *Nature* 323:646–650.

Morton, N. E. (1955) Sequential tests for the detection of linkage. *American Journal of Human Genetics* 7:277–318.

Morton, N. E. (1956) The detection and estimation of linkage between the genes for elliptocytosis and the Rh blood type. *American Journal of Human Genetics* 8:80–96.

Murray, J. C., Nishimura, D., Ardinger, H., Buetow, K., Spence, A., Falk, R., et al. (1988) Linkage of Van der Woude syndrome to markers on chromosome 1q and exclusion of laminin B2 as a candidate gene. *American Journal of Human Genetics* 43 (Suppl.):A153.

Nakamura, Y., Leppert, M., O'Connell, P., Wolff, R., Holm, T., Culver, M., et al. (1987) Variable number tandem repeat (VNTR) markers for human gene mapping. *Science* 237:1616–1622.

Palek, J. (1985) Hereditary elliptocytosis and related disorders. *Clinical Haematology* 14:45.

Rao, D. C., Keats, B. J. B., Morton, N. E., Yee, S. & Lew, R. (1978) Variability of human linkage data. *American Journal of Human Genetics* 30:516–529.

Reeders, S. T., Breuning, M. H., Davies, K. E., Nicholls, R. D., Jarman, A. P., Higgs, D. R., et al. (1985) A highly polymorphic DNA marker linked to adult polycystic disease on chromosome 16. *Nature* 317:542–544.

Riccardi, V. M. & Eichner, J. E. (1986) *Neurofibromatosis: Phenotype, Natural History and Pathogenesis*. Baltimore: Johns Hopkins University Press.

Riordan, J. R., Rommens, J. M., Kerem, B., Alon, N., Rozmahel, R., Grzelczak, Z., et al. (1989) Identification of the cystic fibrosis gene: Cloning and characterization of complementary DNA. *Science* 245:1066–1073.

Risch, N. (1987) Assessing the role of HLA-linked and unlinked determinants of disease. *American Journal of Human Genetics* 40:1–14.

Risch, N. (1990a) Linkage strategies for genetically complex traits. I. Multilocus models. *American Journal of Human Genetics* 46:222–228.

Risch, N. (1990b) Linkage strategies for genetically complex traits. II. The power of affected relative pairs. *American Journal of Human Genetics* 46:229–241.

Risch, N. (1990c) Linkage strategies for genetically complex traits. III. The effect of marker polymorphism on analysis of affected relative pairs. *American Journal of Human Genetics* 46:242–253.

Risch, N., Claus, E. & Giuffra, L. (1989) Linkage and mode of inheritance in complex traits. In *Multipoint Mapping and Linkage Based upon Affected Pedigree Members: Genetics Analysis Workshop 6*, ed. R. C. Elston, M. A. Spence, S. E. Hodge & J. W. MacCluer, pp. 183–188. New York: Alan R. Liss.

Risch, N., Ottman, R. & Shea, S. (1985) Assessing the familial correlation in age at onset: Application to coronary heart disease. *American Journal of Human genetics* 37 (Suppl.):A206.

Rissanen, A. (1979) Familial occurrence of coronary heart disease: Effect of age at diagnosis. *American Journal of Cardiology* 44:60–66.

Robins, L. N., Helzer, J. E., Weissman, M. M., Orvaschel, H., Gruenberg, E., Burke, J. D. & Regier, D. (1984) Lifetime prevalence of specific psychiatric disorders in three sites. *Archives of General Psychiatry* 41:949–958.

Rommens, J. M., Ianuzzi, M. C., Kerem, B., Drumm, M. L., Melmer, G., Dean, M., et al. (1989) Identification of the cystic fibrosis gene: Chromosome walking and jumping. *Science* 245:1059–1065.

Rouleau, G., Wertelicki, W., Haines, J. L., Hobbs, W. J., Trofatter, J. A., Seizinger, B. R., et al. (1987) Genetic linkage of bilateral acoustic neurofibromatosis to a DNA marker on chromosome 22. *Nature* 329:246–248.

Sherrington, R., Brynjolfsson, J., Petursson, H., Potter, M., Dudleston, K., Barraclough, B., et al. (1988) Localization of a susceptibility locus for schizophrenia on chromosome 5. *Nature* 336:164–167.

St. George-Hyslop, P. H., Tanzi, R. E., Polinshy, R. J., Haines, J. L., Nee, L., Watkins, P. C., et al. (1987) The genetic defect causing familial Alzheimer's disease maps on chromosome 21. *Science* 235:885–890.

Weeks, D. E. & Lange, K. (1988) The affected pedigree member method of linkage analysis. *American Journal of Human Genetics* 42:315–326.

Chapter 6

Age, Period, and Cohort Effects on Rates of Mental Disorders

JOHN P. RICE
STEVEN O. MOLDIN
ROSALIND NEUMAN

Introduction

Epidemiological research is concerned with describing and understanding rates of illnesses, disorders, syndromes, and other public health problems in a population, particularly as the rates are affected by characteristics of the individual, place, or time. Temporal or secular trends are variations in rates over time, and are well established for many medical conditions like Parkinson's disease and cardiovascular conditions.

An "age effect" is present for the major mental illnesses. That is, the rates of illness vary with an individual's age. Investigations have traditionally reported age-specific rates and an age-of-onset distribution. Family data have been described using the lifetime morbid risk for a disorder and the morbid risk in classes of relatives of affected individuals. This conceptualization became problematic when data from two major multicenter studies (the Epidemiological Catchment Area Study of the general population and the Collaborative Depression Study of relatives of depressives) found (lifetime) rates in younger individuals to already exceed those in older individuals. Klerman et al. (1985) have discussed several possible methodologic artifacts that might explain observed secular trends—recall, public awareness, different mortality, and the possibility that more recent generations are more "psychologically minded." There is as yet no clear indication whether or not these phenomena could account for the massive effects seen in current studies.

In addition to simple age effects, one can also consider (birth) cohort, where rates at each age are influenced by factors occurring during or around the year of birth. The size of one's birth cohort (the "baby boomers") provides an example that could influence the rates of some disorders. In a cohort effect, rates depend on the year of birth, whereas in a "period effect", the calendar year, which cuts across different cohorts at different ages, is the key determinant. Events such as war, radiation exposure (as in the Three Mile Island accident), or drug availability in the late 1960s are specific to a calendar year. Although it is possible to conceptually distinguish these age, cohort, and period effects (Fienberg & Mason, 1979; Holford, 1983; Lawless, 1982), they are mathematically confounded in a linear relationship. This is further complicated by the observation that more recent birth cohorts, which show increased rates of illness, are frequently observed only partly through the age of risk.

There may be an interaction between two of the three effects. That is, a person may be vulnerable only at certain ages to a calendar event, providing an age-period interaction. Since knowledge of any two of the variables of age, period, and cohort determines the third, however, there is no pure statistical way to disentangle these effects. The two approaches used are that following Holford (1983), which considers estimable, nonlinear relationships, and that following Fienberg and Mason (1979), which considers substantive constraints based on external knowledge.

We first describe some of the statistical methods used to investigate the age/period/cohort (APC) problem, and then describe the recent literature applying these methods to alcoholism and the affective disorders. Finally, we discuss the implications of these findings for genetic research.

Methods Used For Age/Period/Cohort Analysis

The methods used for APC analysis have been typically based either on a log-linear/logistic approach or on a survival analysis approach. As noted below, these two approaches can, in fact, approximate one another by assuming a piecewise exponential model using age intervals.

The Two Basic Statistical Methods

Logistic Regression Model. Consider a dichotomous outcome variable Y, with Y = 1 corresponding to "affected" and Y = 0 corresponding to "unaffected". We wish to consider the effect of a predictor variable X on the probability p of being affected. A natural transformation of p is the logit (log odds) θ,

$$\theta = logit(p) = \log\left(\frac{p}{1-p}\right) = \log\left(\frac{Pr(Y=1)}{Pr(Y=0)}\right),$$

where Pr ( ) denotes the probability of the event in parenthesis. Since p is a probability, it is bounded between 0 and 1, whereas logit (p) has a range from $-\infty$ to $+\infty$. In the logistic model (Fleiss et al., 1986), we assume the logit of p has a linear regression on X, that is,

$$logit (p) = \alpha + \beta X,$$

for some numbers α and β. Exponentiating both sides of the above equation, we have

$$\frac{p}{1-p} = e^{\alpha+\beta X}.$$

The quantity $p/(1-p)$ is called the odds (conditional on X). If X were a dichotomous variable with, for example, X = 1 denoting females and X = 0 denoting males, then the odds of being affected for females would be $e^{\alpha+\beta}$ = $e^{\alpha}e^{\beta}$, and the odds for males, e^{α}. Note that if $\beta = 0$, then $e^{\beta} = 1$, indicating no difference between males and females. Thus, the odds ratio for the variable sex is e^{β}, the exponentiated β. Indeed, if one constructs the usual 2×2 table of sex by affection status, e^{β} is identical to the usual cross product odds ratio in this situation.

The advantage of the logistic approach is that it generalizes to a set of covariates $X' = (X_1, \ldots, X_n)$ as

$$logit (p) = \alpha + \beta_1 X_1 + \ldots + \beta_n X_n.$$

Solving for p, we have

$$p = \frac{e^{\alpha+\beta'X}}{1 + e^{\alpha+\beta'X}}$$

Where $\beta' = (\beta_1, \beta_2, \ldots, \beta_n)$. That is, p (as a function of the covariates) is given by the logistic function.

The covariates $X_1, \ldots, X_n$ may be continuous or discrete variables. One important example is the use of dummy variables in the coding of categorical variables. For example, suppose we have individuals diagnosed as schizophrenic, bipolar, or normal. With three groups, we create two "dummy" independent variables X_1 and X_2 coded as 0 and 1. In this case we may code individuals as

	X_1	X_2	odds
schizophrenic	1	0	$e^{\alpha}e^{\beta 1}$
bipolar	0	1	$e^{\alpha}e^{\beta 2}$
normal	0	0	e^{α}

In this case, $e^{\beta 1}$ is the odds ratio comparing schizophrenics to normals and $e^{\beta 2}$ is the odds ratio comparing bipolars to normals for the dependent variable Y of interest. The statistical significance of each β_i may be tested individually to compare schizophrenics and bipolars to normals.

In general, if a variable has k categories, $k - 1$ dummy variables may be created and $k - 1$ β's introduced into the model. This approach is the one used when age (or period or cohort) categories are created as discussed below.

Survival Analysis. In survival analysis (Kalbfleish & Prentice, 1980) we consider the random variable T, the time to an event such as death or the onset of an illness. Special attention is given to: (1) incomplete (censored) data in which T is not observed for all individuals and (2) covariates which influence the time to the event. The survival function S(t) is the probability of being well (or alive, etc.) at time t,

$$S(t) = 1 - \Pr(T \leq t).$$

The function S(t) may be estimated in a nonparametric way (the Kaplan-Meier estimate) by a "step function", in which S(t) is 1.0 until the time of the first event. The curve will have a drop, based on the number at risk, each time an event happens. Accordingly, if an observation is censored at a certain time, that observation will be used in calculations up to that time in estimating S(t). When data are grouped into discrete time intervals, these techniques are referred to as life-table analysis. The standard statistical packages have routines available for estimating and graphing S(t), and for comparing survival curves from different groups.

Another important function is the hazard function $\lambda(t)$. The hazard at t is the risk of becoming ill at t given that the individual has remained well up to time t. (Technically, it is the density function of T divided by S(t)). One important case is where the hazard function is a constant, $\lambda(t) = \lambda$. Then the survival function is exponential, $S(t) = e^{-\lambda t}$.

Now consider a set of covariates $Z' = (Z_1, \ldots, Z_n)$. The hazard function for an individual with covariate values Z is denoted $\lambda(t;Z)$. The hazard for $Z = 0$, denoted $\lambda_0(t)$ is called the baseline hazard. In the Cox proportional hazards model, covariates are assumed to act multiplicatively on the hazard function, that is

$$\lambda(t;Z) = \lambda_0(t)\, e^{\beta'Z}.$$

Different baseline hazards may be chosen in subsets of the data (or strata). It must be emphasized that the assumption of proportionality is nontrivial. When proportionality for a variable is rejected, a common solution is to stratify according to that variable. It is also possible for covariates to depend on time, although this greatly increases the computational complexity when fitting the Cox model.

The Piecewise Exponential and Logistic Approximation

Following Holford (1983), let the follow-up period be partitioned into mutually exclusive intervals (I) and assume the hazard function λ_i is constant in each time interval. Let Z' denote a set of covariates with regression coefficients β', and assume $\lambda(t;Z) = \lambda_i\, e^{\beta'Z}$ for t in interval i.

Then

$$\log \lambda(t;Z) = \log \lambda_i + \beta'Z,$$

so that analyses may be done using log linear models as in Wickramaratne et al. (1986). Note that since the hazard function is a constant within each interval, the survival function is piecewise exponential.

Another approach is to note that when λ_i is small, $1 - \lambda_i$ is approximately equal to 1, so the following model is approximately the same:

$$\frac{\lambda(t;Z)}{1 - \lambda(t;Z)} = \frac{\lambda_i}{1 - \lambda_i}\, e^{\beta'Z},$$

or

$$\log\left(\frac{\lambda(t;Z)}{1 - \lambda(t;Z)}\right) = \log\left(\frac{\lambda_i}{1 - \lambda_i}\right) + \beta'Z$$

or

$$logit(\lambda(t;Z)) = logit(\lambda_i) + \beta'Z.$$

That is, if we assume the hazard is constant within intervals, it is possible to use a logistic model that will approximate a survival model. This is the approach used in Lavori et al. (1987).

The Identification Problem

The interdependency between age (A), period (P), and cohort (C) gives rise to what is known as the identification problem. This problem is created when attempting to determine the effect of all three factors, A, P, and C on a response variable such as the rates of illness of some disorder. The confounding of these three factors can be illustrated by considering data from a set of age groups and time periods. Assume the ranges of the time periods and age intervals are all equal, that is, if the time periods are divided into 10-year intervals, the range of years for each age group is also 10 years. Birth cohorts can then be determined by knowledge of the age of an individual and the time period when the event of interest occurred. For example, if a subject was in the the 20–29 year age group during the time period 1930–1939 (when the event occurred), then his/her year of birth (the birth cohort) is in the interval 1901–1919. In general, if any two of the three factors, A, P, and C, are specified, the third factor is completely determined.

The confounding of A, P, and C becomes problematic when attempting to decompose the response variable, Ω_{ijk}, as in analysis of variance models: $\Omega_{ijk} = \mu + \alpha_i + \pi_j + \gamma_k$ where $\Sigma_i\, \alpha_i = \Sigma_j\, \pi_j = \Sigma_k\, \gamma_k = 0$. Here α_i, Π_i, and γ_k are the possible age, period, and cohort effect parameters for the ith age group, j^{th} time period, and k^{th} birth cohort, respectively. Because of the interdependency of A, P, and C, not all of the parameters can be uniquely determined. The limitations encountered by the lack of a unique solution to the age/period/cohort problem has stimulated much work in an effort to overcome these complications. We shall briefly outline the solutions offered by two investigators: Fienberg and Mason (1979) and Holford (1983). A full exposition of their work is too complicated to be presented here.

Holford's response variable is $\Omega_{ijk} = \log \lambda_{ijk}$ where λ_{ijk} is the incidence rate for age i, period j, and cohort k. He uses the theory of estimable functions to resolve the identification problem (Searle, 1971). The basic idea is that although the individual effect or parameters cannot be *uniquely* determined, certain linear functions of these parameters may be estimated in the sense that these functions are invariant to any particular solution of the effect parameters. Specifically, Holford reparameterizes each of the effect parameters (the α's, π's, and γ's) into two components: a linear effect and a deviation from linearity or a curvature effect. Using the basic theory of estimable functions, Holford shows that while none of the linear components of age, period, or cohort effects are estimable, any function of the three linear components that has the form $d_1\alpha_L + d_2\pi_L + (d_2 - d_1)\gamma_L$ (where the d's are arbitrary) is estimable. Here α_L, π_L, and γ_L are the linear effects for age, period, and cohort, respectively. Holford also shows that any linear combination of the curvature effects added to the above expression for the linear components is estimable. Wickramaratne et al. (1989) use the method of estimable functions to analyze the effect of age, period, and cohort on major depression. They overcome the limitations of the identification problem by using the fact that the sum of the linear period and linear cohort effects can be estimated. (Let $d_2 = 1$ and $d_1 = 0$ in the above equation.)

Fienberg and Mason (1979) use as their reponse variable $\Omega_{ijk} = \log(P_{ijk}/(1 - P_{ijk}))$, where P_{ijk} is the probability of a positive response for an individual in age group i, period j, and cohort k. They propose an alternative approach to the identification problem, suggesting that external constraints be placed on the effect parameters in order to reduce the number of independent parameters to be estimated, thereby identifying the model. They refer to these constraints as "identification specifications". These restrictions on the parameters may include setting the effect of being in age category i to a constant (α_i = constant) or constraining the effect of being in one cohort equal to the effect of being in another cohort ($\gamma_i = \gamma_j$, i ≠ j). For example, Lavori and colleagues (1987) equated the age effects in two age brackets (20–24 and 25–29) in order to be able to identify the pure effects. They based their assumption on clinical judgment that rates of depression vary little across the decade of age 20–29.

We have presented two methods currently in use to overcome the limitations of age, period, cohort analysis on rates of illness. Each of these methods requires the researcher to make an assumption about his/her data. Fienberg and Mason (1979) caution that the identification specification should not be arbitrary but should include " . . . prior expectations about the nature of at least some of their effects" (p. 61).

Evidence For Secular Trends in Psychiatric Disorders

In several large family studies of psychiatric illnesses reported in the recent literature, evidence has accumulated that supports the hypothesis of important secular changes in rates of the affective disorders and alcoholism. Since secular effects can be sensitive indicators of changing environment, they can (1) provide evidence regarding the role of exogenous factors effecting the onset of the common familial psychiatric conditions, and (2) identify cohorts of individuals who are at increased risk for illness. Therefore, identification and delineation of secular trends have important implications for understanding the etiology of psychiatric illness and for clinical practice.

Alcoholism

Cloninger and colleagues (1988), in a blind interview family study of alcoholic probands and their first-degree relatives, found that the cumulative probability of definite alcoholism increased markedly in those born more recently. Figure 6.1(a) shows that the cumulative risk by age 25 increases with the year of birth from 26% in male relatives born before 1924, 34% for the 1925–1934 birth cohort, 52% for the 1935–1944 birth cohort, 63% for the 1945–

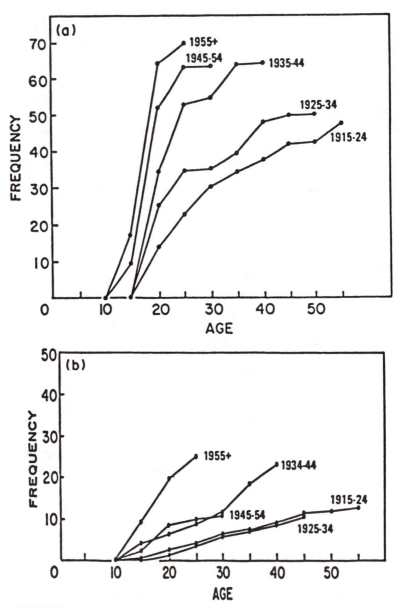

FIGURE 6.1.
Cumulative probability of alcoholism in male (a) and female (b) relatives by age and birth cohort. Source: Reprinted, by permission, from Cloninger et al. (1988). Copyright 1988 by Raven Press

1954 cohort, and 67% for relatives born after 1954. Men in the younger birth cohorts have yet to pass through the full period of risk: therefore, these high nonadjusted risks are especially striking. The cumulative probability of definite alcoholism in the female first-degree relatives is similarly plotted in Figure 6.1(b). Although some adjacent birth cohorts overlap, the more recent cohorts have higher risks overall. In the female relatives of alcoholics the risks at age 25 increase from 3% in women born before 1935, 9% in women born from 1935 to 1954, and 25% in women born after 1954.

These risks in first-degree relatives of alcoholics can be compared with cumulative lifetime prevalence of alcohol abuse/dependence diagnosed according to DIS/DSM-III criteria (Robins et al., 1981) in men and women in the Epidemiological Catchment Area Study (ECA: Robins et al., 1984). Data from the St. Louis sample show that rates for subjects in the following age groups are as follows: 18–24, 17%; 25–44, 21%; 45–64, 11.7%; and 65+, 7.2%. Analysis of the St. Louis ECA data by Cloninger et al. (1988) showed that the lifetime risk of alcoholism in the general population increased for those born since 1953 (males, 22.2%; females, 9.5%) compared with those born before 1938 (males, 12.9%; females, 4%). Again, it is striking that younger subjects have higher risks than do older subjects even though the former have been at risk for shorter periods. These observations are further evidence that traditional methods of age correction are not applicable to mental disorders with such temporal trends.

In a recent analysis of data on alcohol-related problems from four general population surveys, Hasin and colleagues (1990) showed that the prevalence of a "multiple alcohol problem" condition has increased over time. In a comparison of data from the 1967 and 1984 surveys, lifetime rates approximately doubled in females and showed increases of approximately 50% in males. Although psychiatric diagnoses were not determined from the survey data, these results are consistent with those reported above in supporting the hypothesis of secular changes in the rates of alcoholism and increased rates among recent birth cohorts.

Bipolar Illness

Gershon and colleagues (1987) used survival analysis of the cumulative hazard of illness in different birth cohorts and found an increase since 1940 in the rate of the manic spectrum of disorders (schizoaffective, bipolar I, bipolar II) in relatives of bipolar and schizoaffective patients. The authors pooled bipolar and schizoaffective patients and relatives based on family studies and other evidence (reviewed in Nurnberger and Gershon, 1984). The cumulative hazard was greatest for cohorts born since 1940, and there appeared to be a progressive increase in cumulative hazard for successive birth cohorts starting with the earliest-observed cohort.

The authors report that while the differences among the hazard functions were nonsignificant, pooling of relatives into a pre-1940 and a post-1940 cohort reveals a striking difference in hazard functions (Figure 6.2). The cohort born between 1930 and 1939 has a lower cumulative hazard of mania compared with the cohort born a decade earlier. For the rest of the cohorts, the cumulative hazard at the last ages observed is higher than the value for the cohort born in the previous decade at the same age. The possibility that the post-1940 cohorts have accelerated age-of-onset

FIGURE 6.2.
Bipolar and schizoaffective illnesses in relatives pooled into pre-1940 and post-1940 cohorts. Source: Reprinted, by permission, from Gershon et al. (1987)

with compensatory decrements of onsets later in life cannot be ruled out from these data since the cohorts have not been followed through the entire period of risk. Examination of Figure 6.2, in fact, suggests that lifetime morbid risk is the same with an accelerated onset in the more recently born individuals.

Rice and colleagues (1987) analyzed data from the NIMH Collaborative Study on the Psychobiology of Depression and found that age-of-onset to bipolar disorder was accelerated with birth cohort, with individuals born in more recent cohorts having an earlier onset. Using the Cox proportional hazards model, the authors found that both the proband's age-of-onset and relative's birth cohort were significantly related to risk of illness in relatives. Figure 6.3 gives Kaplan-Meier estimates of the risk for relatives in three age groups: 25 and under, 26–44, and 45+. These curves show a dramatic effect for birth cohort; for example, the youngest onset in individuals 45 years and older is reported to be at age 20, whereas half of the onset in the middle cohort occur at this age. The authors note their evidence for birth cohort effects must be interpreted cautiously, with the accelerated onsets for bipolar illness reflecting either younger susceptible individuals expressing their phenotype at an earlier age ("true" accelerated age-of-onset) or a period effect because relatives were interviewed over a narrow time interval. Longitudinal assessment and follow-up are required to permit distinction between these two alternatives. Weissman et al. (1988) analyzed ECA data and presented the following one-year prevalence rates for bipolar disorder in the corresponding age groups: 18–44, 1.4%; 45–64, 0.4%; 65+, 0.1%. These results are compatible with the evidence from family studies in favor of important secular trends for bipolar disorder.

Unipolar Depressive Illness

The most consistent evidence supporting secular trends for a mental disorder exists in regard to unipolar depression. Klerman and Weissman (1989) reviewed findings of several recent, large epidemiologic and family studies that suggest an increase in the rates of illness in the cohorts born after World War II; an increase between 1960 and 1975 in the rates of depression for all ages; and a decrease in the age of onset with an increase in the late teenage and early adult years. Several studies using comparable methods

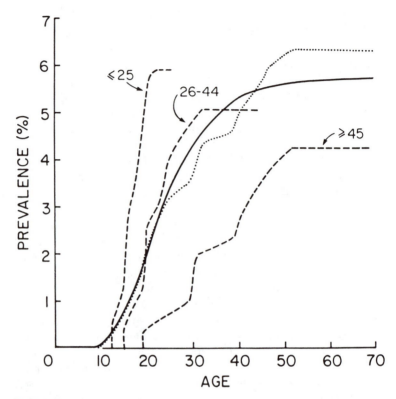

FIGURE 6.3.
Risk of bipolar illness by age for the Stromgren method utilizing ages at onset in probands and computing morbid risk in relatives (smooth curve); survival analysis (Kaplan-Meier) estimate using first-degree relatives (dotted curve); and Kaplan-Meier estimates within indicated three age strata (dashed curves). Source: Reprinted, by permission, from Rice et al. (1987). Copyright 1987 by the American Medical Association

and modern diagnostic criteria have been conducted in the United States, Sweden, Germany, Canada, New Zealand, Korea, and Puerto Rico and are discussed in detail in Klerman and Weissman (1989). In this review, we will focus on the specific findings from the Lundby study (Hagnell et al., 1982), the Collaborative Study on the Psychobiology of Depression (Klerman et al., 1985; Lavori et al., 1987; Rice et al., 1984), the ECA study

(Robins et al., 1984; Weissman et al., 1988; Wickramaratne et al., 1989), and the Camberwell Registry study (Sturt et al., 1984).

The Lundby study consists of all inhabitants in a delimited area in Sweden initially studied by Essen-Moller and colleagues (1956). At the 10-year follow-up in 1957 (Essen-Moller & Hagnell, 1961) there was already an increased number of cases of unipolar depression compared with that expected according to the 1947 point-prevalence figures. Comparisons over a 25-year follow-up (Hagnell et al., 1982) were made of the incidence figures for uni-polar depression during the periods 1947–1957 and 1957–1972, and during the 5-year intervals within the period. The probability of having unipolar depression increased over the 25-year period for both sexes, with a considerably increased (10-fold) probability of depression among young adult men.

Rice et al. (1984) used survival analysis to assess the effects of age and sex on the rates of unipolar depression in relatives in the Collaborative Study. The results showed that when the data were stratified according to sex and birth cohort ($\leq$ 25, 26–44, $\geq$ 45), higher rates of illness were observed in the younger cohorts. The magnitude of female-male differences varied across cohorts, with the least female-male difference (1.6) being found in the youngest cohort.

A detailed application of life-table methodology to this family data set was presented in Klerman et al. (1985). Figure 6.4 shows the cumulative probability for male and female relatives at each age when an episode of unipolar depression occurred. These curves depict progressively increasing rates of unipolar depres-sion in successive birth cohorts through the twentieth century, with higher rates for female subjects in all birth cohorts. The data also indicated a progressively earlier age at onset for male and female subjects in each birth cohort.

Lavori and colleagues (1987), using data only from siblings un-der age 50 in the Collaborative Study, applied Cox proportional hazard models and other survival regression methods to test the statistical significance of the effects of cohort of birth and year of interview on risk to illness. Results of their age/period/cohort analysis suggested a powerful period effect may be responsible for the secular trends in rates of unipolar depression, with rates of onset for siblings between 15 and 50 years doubling between the 1960s and 1970s. This conclusion was based on two assump-tions: the trend (period or cohort) is real and is not an artifact of

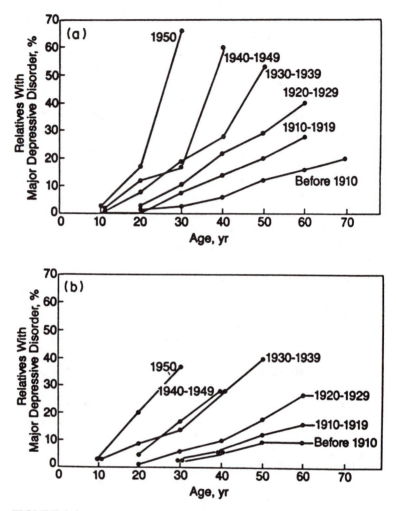

FIGURE 6.4.
Cumulative probability of diagnosable major depressive
disorder in female (a) and male (b) relatives by birth cohort.
Source: Reprinted, by permission, from Klerman et al. (1985).
Copyright 1985 by the American Medical Association

forgetting, and the age-specific rates of risk are constant across the ages 20–29.

Lifetime prevalence of DSM-III major depressive episode was reported in Robins et al. (1984) in three of the sites in the ECA study. Generally, a higher rate of illness was reported in younger cohorts than older ones. For example, in New Haven the rates of illness for different birth cohorts with males and females combined were as follows: 18–24, 7.5%; 25–44, 10.4%; 45–64, 4.2%; and 65+, 1.8%. Weissman et al. (1988) reported one-year prevalence of DIS/DSM-III major depression across five sites in the ECA study. The results show a nearly fourfold increase of illness in women, with higher rates for women and increasing rates in younger cohorts for both males and females (males: 16–44, 1.6%; 45–64, 1.3%; 65+, 0.4%—females: 18–44, 4.8%; 45–64, 2.9%; 65+, 1.4%).

Wickramaratne et al. (1989) analyzed the ECA data using Holford's (1983) methods and also reported strong cohort and period effects. Their results showed a sharp increase in rates of unipolar depression among both men and women in the birth cohort born during 1935–1945. The rates associated with period of onset of unipolar depression continued to increase between the years 1960 and 1980 among both men and women of all ages studied. Figure 6.5(a) and (b) shows cumulative rates of illness for both sexes across all five ECA sites. These plots reveal a trend for the cumulative risk of illness to increase with each successive birth cohort. The cumulative risk among females is consistently higher than among males for all birth cohorts considered, but the male-female differences seem to be decreasing among the more recent birth cohorts.

Sturt et al. (1984) analyzed records in the Camberwell (London, England) Register and found a decline in the rate of unipolar depression during the 1970s. Lifetime morbid risk to depression for the years 1971, 1976, and 1981 were as follows: males: 10.0%, 11.9%, 9.4%; females: 18.9%, 20.2%, 12.3%. This study is unique in reporting contrary evidence in favor of a secular trend of *decreasing* rates of depressive illness with younger cohorts. However, the data gathered by Sturt et al. (1984) are based on clinical diagnoses from the Registry and therefore are not directly comparable to data from the other studies we reviewed above, in which diagnoses were derived using face-to-face interviews with standardized instruments.

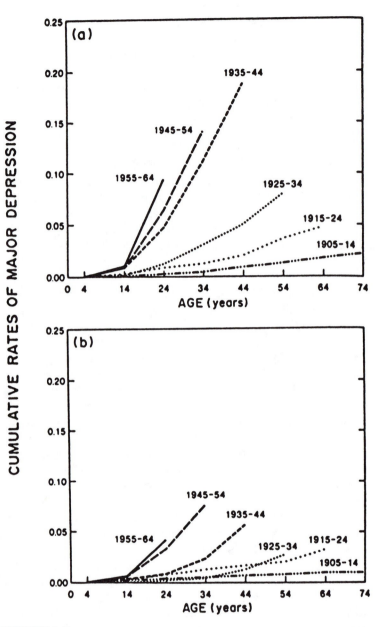

FIGURE 6.5.
Cumulative rates of major depression in females (a) and males (b) by birth cohort—pooled five-site data. Source: Reprinted, by permission, from Wickramaratne et al. (1989). Copyright 1989 by Pergamon Press, Inc.

Implications For Genetic Models

Secular changes over one generation may have a major impact on understanding the genetics of mental illness. Even if the sources of observed trends turn out to be due to methodologic artifacts (although this has not been demonstrated), the effect is so prominent that it must be incorporated into analytic strategies. It must be emphasized that such a change over one generation is not due to genetic changes in the population (e.g., changes in gene frequencies), but must reflect environmental changes. However, this does not mean that genes are not involved or that the magnitude of genetic influences has necessarily changed.

To illustrate this last point, we give an example used by animal breeders. Consider a herd of cattle maintained on a uniformly poor diet. If the variation in adult weight about the mean is controlled only by many genes, the character "adult weight" is 100% heritable. If the diet is changed to a uniformly rich one, there can be a massive shift in the mean adult weight due to the environmental change. Adult weight in this new environment may again be 100% heritable, or have reduced heritability. That is, genetic effects refer to variation in the population, and a change in environment may give rise to a mean shift which may or may not influence the sources of variation in the population.

Genetic models for psychiatric phenotypes are usually based on an underlying liability distribution with thresholds; an individual is "affected" if his liability is above the threshold (Rice & McGuffin, 1985; Rice et al., in press). Thus, a mean shift in liability can move a greater proportion of the population above the threshold with a corresponding increase in the prevalence. In this case, the model for the mode of transmission would remain the same, with only the prevalence changing. Alternatively, greater environmental variance could be introduced, or the genes may have a greater or lesser effect in the new environment. Accordingly, the presence of secular trends per se does not imply that the mode of transmission has changed or that the genetics of the disorder has become more complicated (the penetrance of certain susceptible individuals may have been increased and a gene may be easier to detect).

What is clear is that the modeling and analysis of family data are more complex in the presence of secular trends. It is no longer sufficient to use only an individual's age at observation to correct for censoring; cohort and period may also enter the calculations. Whether the trend is due to age, period, or cohort, or to an interactive effect of, for example, age by period, will make a difference in the fitting of genetic models. Follow-up data in families will ultimately be needed to verify trends in more recently born cohorts.

Discussion

The recognition of secular trends for mental disorder is a phenomenon of the last decade. for drug abuse disorder, the wide availability of street drugs starting in the late 1960s is a natural candidate for a period effect for this disorder. In fact, there appears to be an age-period interaction with only the young being influenced. For the other disorders the source of reported secular trends is speculative.

Rice et al. (1987) attributed the trends in bipolar illness to an acceleration of the age of onset, rather than to a marked change in cumulative incidence by birth cohort. The data of Gershon et al. (our Figure 6.2) could also be given that interpretation. The distinction between these hypotheses involves extrapolation at the end of the survival curves for more recently born cohorts, so that follow-up data will be useful.

For major depression, the observed trends cannot be solely explained by an accelerated onset. The λ's in the Cox proportional hazards model from the Collaborative Depression Study and the ECA study are both 0.07 when birth cohort is entered into the Cox proportional hazards model. The odds/year are then 1.07, so the effect over a decade would be $e^{.7} = 2$. Lavori et al. (1987) restricted their analyses to those under age 53, so that their finding does not result from a low rate in the elderly, which could be explained by recall effects. They also comment that the secular trend remains if more severe criteria are used for diagnosis.

A major question to be resolved is whether the degree of familial resemblance for these disorders is affected by recent trends. High rates of affective disorders have been reported in children at risk (Welner & Rice, 1988), and further work is needed to separate the familial and secular factors. For alcoholism, Reich et al. (1988) found sex- and cohort-specific transmissibilities, with greater transmissibility in the youngest cohorts.

A related question is whether the increased number of cases in more recently born cohorts reflects an increase in heterogeneity in these disorders. This heterogeneity could result from individuals who express a different genetic susceptibility or who represent a form of illness due to environmental etiology. Geneticists distinguish between cases and phenocopies, individuals who have the same phenotype but for nongenetic causes. Phenocopies (or false positives) present complications for methods such as linkage analysis, so that observed secular trends are important not only for proper "age correction," but for any modern genetic study.

Acknowledgments

Supported in part by USPHS grants MH-37685, MH-31302, MH-43028, MH-25430, AA08028, NIMH Research Training Grant MH-14677 (RN and SOM), and the MacArthur Network I Task Force on Analytic Strategies for Linkage Analysis of Psychiatric Disorders.

References

Cloninger, C. R., Reich, T., Sigvardsson, S., von Knorring, A.-L. & Bohman, M. (1988) The effects of changes in alcohol use between generations on the inheritance of alcohol use. In *Alcoholism: Origins and Outcomes*, ed. R. M. Rose & J. E. Barrett, pp. 49–74. New York: Raven Press.

Essen-Moller, E. & Hagnell, O. (1961) The frequency and risk of depression within a rural population group in Scania. *Acta Psychiatrica Scandinavica* 37 (Suppl. 162):28–32.

Essen-Moller, E., Larsson, H., Uddenberg, C. E. & White, G. (1956) Individual traits and morbidity in a Swedish rural population. *Acta Psychiatria et Neurologica Scandinavica* 100 (Suppl.).

Fienberg, S. E. & Mason, W. (1979) Identification and estimation of age-period-cohort models in the analysis of discrete archival data. In *Sociological Methodology*, ed. Karl F. Schuessler. San Francisco: Jossey-Bass.

Fleiss, J. L., Williams, J. B. W. & Dubro, A. F. (1986) The logistic regression analysis of psychiatric data. *Journal of Psychiatric Research* 20:145–209.

Gershon, E. S., Hamovit, J. H., Guroff, J. J. & Nurnberger, J. I. (1987)

114   J. P. RICE, S. O. MOLDIN, R. NEUMAN

Birth-cohort changes in manic and depressive disorders in relatives of bipolar and schizoaffective patients. *Archives of General Psychiatry* 44:314–319.

Hagnell, O., Lanke, J., Rorsman, B. & Ojesjo, L. (1982) Are we entering an age of melancholy? Depressive illnesses in a prospective epidemiological study over 25 years: The Lundby Study, Sweden. *Psychological Medicine* 12:279–289.

Hasin, D., Grant, B., Harford, T., Hilton, M. & Endicott, J. (1990) Multiple alcohol-related problems in the U.S.: On the rise? *Journal of Studies on Alcohol* 51:485–493.

Holford, T. R. (1983) The estimation of age, period, and cohort effects for vital rates. *Biometrics* 39:311–324.

Kalbfleish, J. D. & Prentice, R. L. (1980) *The Statistical Analysis of Failure Time Data*. New York: John Wiley.

Klerman, G. L., Lavori, P. W., Rice, J. P., Reich, T., Endicott, J., Andreasen, N. C., et al. (1985) Birth-cohort trends in rates of major depressive disorder among relatives of patients with affective disorder. *Archives of General Psychiatry* 42:689–693.

Klerman, G. L. & Weissman, M. M. (1989) Increasing rates of depression. *Journal of the American Medical association* 261:2229–2235.

Lavori, P. W., Klerman, G. L., Keller, M. B., Reich, T., Rice, J. & Endicott, J. (1987) Age-period-cohort analysis of secular trends in onset of major depression: Findings in siblings of patients with major affective disorder. *Journal of Psychiatric Research* 21:23–35.

Lawless, J. F. (1982) *Statistical Models and Methods for Lifetime Data*. New York: John Wiley.

Nurnberger, J. I. & Gershon, E. S. (1984) Genetics of affective disorders. In *Neurobiology of Mood Disorders*, ed. R. Post & J. Ballenger, pp. 76–101. Baltimore: Williams & Wilkins.

Reich, T., Cloninger, C. R., Van Eerdewegh, P., Rice, J. P. & Mullaney, J. (1988) Secular trends in the familial transmission of alcoholism. *Alcoholism: Clinical and Experimental Research* 12:458–464.

Rice, J. & McGuffin, P. (1985) Genetic etiology of schizophrenia and affective disorder. In *Psychiatry*, vol. 1, ed. R. Michels & J. O. Cavenar, Jr., et al., pp. 1–24. Philadelphia: Lippincott.

Rice, J., Neuman, R. & Moldin, S. O. (in press) Methods for the inheritance of qualitative traits. In *Handbook of Statistics—VIII. Statistical Methods in Biological and Medical Science*, ed. C. R. Rao & R. Chakraborty. North-Holland.

Rice, J., Reich, T., Andreasen, N. C., Endicott, J., Van Eerdewegh, M., Fishman, R., et al. (1987) The familial transmission of bipolar illness. *Archives of General Psychiatry* 44:441–447.

Rice, J., Reich, T., Andreasen, N. C., Lavori, P. W., Endicott, J., Clayton, P. J., et al. (1984) Sex-related differences in depression: Familial evidence. *Journal of Affective Disorders* 71:199–210.

Robins, L. N., Helzer, J. E., Croughan, J. L. & Ratcliff, K. S. (1981) The NIMH Diagnostic Interview Schedule: its history, characteristics, and validity. *Archives of General Psychiatry* 38:381–389.

Robins, L. N., Helzer, J. E., Weissman, M. M., Orvaschel, H., Gruenberg, E., Burke, J. D. & Regier, D. A. (1984) Lifetime prevalence of specific psychiatric disorders in three sites. *Archives of General Psychiatry* 41:949–958.

Searle, S. R. (1971) *Linear Models*. New York: Wiley.

Sturt, E., Kumakura, N. & Der, G. (1984) How depressing life is—life-long morbidity risk for depressive disorder in the general population. *Journal of Affective Disorders* 7:109–122.

Weissman, M. M., Leaf, P. J., Tischler, G. L., Blazer, D. G., Karno, M., Bruce, M. L. & Florio, L. P. (1988) Affective disorders in five United States communities. *Psychological Medicine* 18:141–153.

Welner, Z. & Rice, J. (1988) School-aged children of depressed parents: A blind and controlled study. *Journal of Affective Disorders* 15:291–302.

Wickramaratne, P. J., Prusoll, B. A, Merikangas, K. R. & Weissman, M. (1986) The use of survival time models with nonproportional hazard functions to investigate age of onset in family studies. *Journal of Chronic Diseases* 39:389–397.

Wickramaratne, P. J., Weissman, M. M., Leaf, P. J. & Holford, T. R. (1989) Age, period and cohort effects on the risk of major depression: Results from five United States communities. *Journal of Clinical Epidemiology* 42:333–343.

Part 2

Reviews of Substantive Findings

Chapter 7
The Genetics
of Schizophrenia

MICHAEL J. LYONS
KENNETH S. KENDLER
ANNE GERSONY PROVET
MING T. TSUANG

There are three basic genetic questions about schizophrenia that are important to answer: (1) Is something inherited? (2) If yes, what is inherited? and (3) How is it inherited? This chapter will be organized according to these basic questions. In addressing them, many of the important issues about the genetics of schizophrenia will be raised.

Is Something Inherited?

Answers to this question began to be provided shortly after the modern formulation of schizophrenia was developed by Kraepelin (1971). Initially, Kraepelin, Bleuler, and other clinicians noticed the presence of certain types of "schizophrenia-like" symptoms among the relatives of their schizophrenic patients. There were also more rigorous efforts to adduce evidence of familial aggregation. Rudin, a pupil of Kraepelin, began the formal investigation of the familial aspects of schizophrenia with a family study of dementia praecox in 1916.

Rudin's early study ushered in what could be called the "first generation" of familial aggregation studies, primarily conducted in Europe, that provided evidence for familial transmission. The consistent finding from these studies was that relatives of schizophrenics had a greater risk for schizophrenia than that observed in the general population. Gottesman and Shields (1982) reviewed the familial aggregation studies through 1980. They summarized the accumulated data on the morbid risk (MR) (morbid risk refers to the probability that an individual will develop schizophrenia at some time if he or she lives through the entire risk period for the illness) for definite plus probable schizophrenia to various types of relatives. Among first-degree relatives (those sharing 50% of their genes with the proband), parents had an average MR of 5.6%, siblings had an average MR of 10.1%, and offspring an average MR of 12.8%. The lower risk to parents compared to other first-degree relatives is due to the effects of schizophrenia on fertility. Specifically, schizophrenia decreases the probability that one will reproduce. In a sample selected for having reproduced (i.e., parents), there is a bias toward lower rates of schizophrenia. There is no such selective factor operating among the siblings and offspring of schizophrenic probands. The MR for the various classes of second-degree relatives (those sharing 25% of their genes with the proband) ranges from 2.4% for uncles and aunts to 4.2% for half siblings. The MR among first cousins, one type of third-degree relative, was 2.4%.

Kendler (1988a) reviewed familial aggregation studies and identified a number of methodological shortcomings that were widespread in the first generation of studies. One important limitation was the absence of control or comparison groups. In many studies the MR to relatives was determined and it was compared to rates observed from other studies of the general population. Thus, there were no comparisons among groups within the same study, making it difficult to interpret results. Another serious problem was the lack of "blindness" in the assignment of diagnoses. That is, in many cases the assignment of diagnoses to relatives was made with the diagnosticians aware of the probands' diagnoses. This introduces the possibility that the diagnosis of relatives could be influenced by the diagnosis of the probands. (This does not imply wrongdoing on the part of the investigators, but diagnosticians may be influenced in their judgments by their expectations.) Finally, many of these studies were problematic with regard to how the diagnoses were made. The modern standard for research diagnoses includes data collection through structured diagnostic interviews and the application of standardized, explicit diagnostic criteria. Many of the first generation of family studies did not specify how data were collected, whether through interview, informants, or records such as medical charts. Frequently, the method for arriving at a diagnosis was not made explicit.

Kendler (1988a; 1988b) also reviewed a number of family studies conducted since 1980 that are representative of a "second generation" of studies that avoided the methodological short-comings that plagued the first generation. These studies included a comparison group, structured diagnostic interviews, and the assurance of blindness in the diagnostic process. Table 7.1 summarizes the findings from his review.

These studies found MRs to first-degree relatives ranging from 8.9% (Scharfetter & Nusperli, 1980) to 1.4% (Coryell & Zimmerman, 1988). Although the absolute values of MRs were considerably lower than those in the first generation of studies, reflecting narrower diagnostic criteria, six of the eight second-generation studies demonstrated a statistically significant difference between the relatives of schizophrenics and those of controls and another showed a trend in the same direction. This underscores the importance of including a comparison group because the modern, narrower definition of schizophrenia results in lower rates in both the relatives of schizophrenics and those of controls, leaving the relative risk for schizophrenics' relatives quite similar to that observed in the older studies.

In addition to Coryell and Zimmerman (1988), there have been two other recent studies (Abrams & Taylor, 1983; Pope et al., 1982) that failed to replicate findings of an elevated MR for schizophrenia among the relatives of schizophrenic probands. However, these studies have been criticized on methodological grounds. Coryell and Zimmerman (1988) used relatively small samples of relatives, resulting in low statistical power. Their reliance on a high proportion of telephone interviews was questioned because some subtle aspects of schizophrenia-related behavior may best be recognized face to face. They were also criticized for potential unrepresentativeness of their comparison group (Kendler, 1988a). Pope et al. (1982) and Abrams and Taylor (1983) were criticized because neither study used a control group and because one depended exclusively on the family history method (Pope et al., 1982) and the other assessed most relatives with that method. (The *family history* method refers to obtaining information about relatives from an informant or records, whereas the *family study* method refers to obtaining diagnostic information directly from the subject. Lower MRs are likely to be obtained using the family history method because it is less sensitive in detecting schizophrenia than the family study method (Andreasen et al., 1977).)

Because the preponderance of strong evidence supports the

TABLE 7.1.
Summary of the "Second Generation" of Published Familial Aggregation Studies of Schizophrenia That Meet Strict Methodological Criteria

Source	Criteria	Controls	Different groups of, or definitions of, illness in relatives	Relatives of schizophrenic probands			Relatives of control probands		
				BZ^a	Schizophrenia N of cases	MR (%)	BZ^a	Schizophrenia N of cases	MR (%)
Scharfetter & Nusperli (1980)	ICD-9	Affective illness	—	550	49	8.9	451	15	3.3**
Tsuang et al. (1980)	Consensus senior Iowa clinicians	Screened surgical patients	All evaluated relatives	362	20	5.5	475	3	0.6**
			Only interviewed relatives	342	11	3.2	473	3	0.6*
Guze et al. (1983)	Modified Washington University	Nonschizophrenic psychiatric patients	Definite schizophrenia	111	4	3.6	1,076	6	0.6**
			Definite and probable schizophrenia	111	9	8.1	1,076	18	1.7**
Baron et al. (1985)	RDC, DSM-III	Normal subjects	—	329	19	5.8	337	2	0.6**

Study	Criteria	Control group		N	n	%[a]	N	n	%[a]
Kendler et al. (1985)	DSM-III	Screened surgical patients	All evaluated relatives	703	26	3.7	931	2	0.2**
			Only interviewed relatives	653	12	1.8	919	0	0**
Frangos et al. (1985)	DSM-III	Normal subjects	Definite schizophrenia	478	19	4.0	536	4	0.7**
			Definite and probable schizophrenia	478	26	5.4	536	6	1.1**
Coryell & Zimmerman (1988)	RDC	Never-ill volunteers	—	72	1	1.4	160	0	0
Gershon et al. (1988)[b]	RDC	Normal subjects	—	97	3	3.1	349	2	0.6[c]

SOURCE: Adapted from Kendler (1988)

[a]BZ indicates bezugsziffer (lifetimes at risk)

[b]probands with chronic schizophrenia

[c]$p < .08$

*$p < .01$ **$p < .001$

familial aggregation of schizophrenia, we can answer the question whether something about schizophrenia is inherited in the affirmative. A corollary of this question is whether the "inherited aspect" of schizophrenia is genetic or cultural. There are two broad mechanisms by which the environment might influence the probability of the cooccurrence of schizophrenia in members of the same family—the shared environment and vertical cultural transmission (VCT). In mammalian species the offspring share an environment produced, in part, by their parents. The offspring are likely to share exposure to toxic or infectious agents that could lead to similar outcomes for the siblings. Kendler (1988c) described two subtypes of VCT. Direct vertical cultural transmission refers to a process in which the presence of schizophrenia in the parent increases the likelihood of schizophrenia in the offspring through some type of learning, such as modeling. Indirect VCT refers to a process in which the parents are not themselves schizophrenic, but provide a familial environment that promotes the development of schizophrenia in the offspring (schizophrenogenic parents). In the traditional familial aggregation study, the effects of genetic and familial environmental factors are confounded.

Adoption Studies

The role of adoption studies can be seen as the delineation of the relative contribution of genetic and environmental factors in the transmission of schizophrenia. Heston (1966) studied 47 children born to schizophrenic mothers and 50 controls who were removed from their biological mothers soon after birth and reared by normal adoptive parents. He found a 16.6% morbid risk of schizophrenia in the former group as opposed to a 0% morbid risk in the controls. He further noted that the rate for schizophrenia in children adopted away from their affected parent was close to that of children who were reared by their natural schizophrenic parent. Wender et al. (1974) employed a cross-fostering design, and found a 10.7% prevalence of schizophrenia in children of normal parents who were reared by a schizophrenic parent, compared with 18.8% for children of parents with certain or probable acute, borderline, or chronic schizophrenia raised in normal environments and 10.1% in the adoptive controls. The results appear to indicate that the environment provided by a schizophrenic parent is ineffective in producing schizophrenia.

It must be noted, however, that although numerically higher, the prevalence for children with schizophrenic genotypes is not significantly higher than that of the cross-fostered group. Gottesman and Shields (1982) therefore concluded that these results "do not directly strengthen a genetic argument so much as they weaken environmental ones" (p. 138).

Kety et al. (1968) followed schizophrenic adoptees as probands and examined the prevalence of schizophrenia in their biological and adoptive parents. For the adoptee probands, rates of 12.1% and 1.6% were observed for biological and adoptive parents respectively; 6.2% of the biological parents of normal control adoptees had schizophrenia, compared with a prevalence of 4.4% for adoptive parents. These results are consistent with a genetic theory of transmission. Gottesman (1978) concluded that the "major accomplishment of the adoption studies was to rule out some environmental factors as either necessary or sufficient for the occurrence of schizophrenia" (p. 64). Reveley et al. (1987) concluded that adoption research "supports the contention that the familial aggregation of schizophrenia is largely an expression of genetic factors" (p. 75). Both conclusions support an interpretation of adoption studies as emphasizing genetic factors in the familial transmission of schizophrenia.

Twin Studies

The twin method, like the adoption method, offers another strategy for disentangling the influence of genetic factors and the family environment. The twin method capitalizes on an experiment provided by nature—monozygotic twins (MZ) share all of their genes, whereas dizygotic twins (DZ), like other siblings, share, on average, 50% of their genes. In the classical twin design, both types of twins share the same family environment. (The chapter by Segal et al. reports on the twins reared apart method.) The influence of the family environment is assumed to be the same for MZs and DZs, whereas the shared genetic influence for DZs is one half of that for MZs. To the extent that a phenotype is due to the family environment, the resemblance between DZ twins should be the same as that between MZ twins. For example, for language spoken, there is no difference in resemblance between DZ and MZ twins. For a trait under strong genetic influence, the genetically identical MZ twins will resemble one another more than will DZ twins. For example, MZs will resemble one another more closely for eye color than will DZs. (See the chapter by Rose for a detailed description of the twin method.)

An important issue for determining the appropriateness of the twin method for studying a particular trait is whether the trait is related to twinning per se. For example, autism is more common among twins, therefore, findings from twin studies of autism may not be generalizable to nontwins (Gottesman & Shields, 1982). There is evidence that schizophrenia is neither more nor less common among twins compared to singletons (Allen & Pollin, 1970; Rosenthal, 1960), which suggests that findings from twin studies of schizophrenia are generalizable. Another methodological concern is whether MZ twins have an environment that is more similar than that for DZ twins. If this were the case and MZ twins were more similar in *both* their genes and their environment, the ability of the twin method to unconfound genetic and environmental influences could be jeopardized. A number of studies that examined nonpsychiatric phenotypes have concluded that, to the extent that MZ twins have more similar environments than same-sex DZ twins, the greater MZ environmental similarity is the result of the greater phenotypic similarity of MZ twins rather than the cause of it (Lytton, 1977; Matheny, 1979). Also, to sustain this type of objection to twin studies of schizophrenia, it would be necessary to assert that the features of the environment that are etiologically significant for schizophrenia are more similar for MZ than DZ twins. Given the scant evidence for the role of vertical cultural transmission in schizophrenia, it would seem that the burden of proof rests with critics of the twin method to demonstrate that "trait-relevant" environmental factors are more similar for MZ than same-sex DZ twins. Kendler (1983) reviewed the potential sources of bias in the twin method and concluded that it is unlikely that the twin method will lead to biased results in studies of schizophrenia.

The basic data provided by twin studies of a categorical trait, such as schizophrenia, are the concordance rates observed among MZ and DZ twin pairs. Concordance rate refers to the percentage of twin pairs in which both members are affected with the characteristic of interest. In general, there are two types of concordance rates that are reported in twin studies. The pairwise concordance rate is calculated by dividing the number of pairs in which both members are affected by the total number of pairs with any affected member. It is appropriate to use the pairwise rate when the source of twins is an unselected population such as a general twin register. When the twins are selected or ascertained because they are affected, then the probandwise rate is the appropriate measure of concordance. Among pairs concordant for the characteristic, some will be identified twice—once for each affected twin. The probandwise rate corrects for this "double ascertainment" of pairs. The calculation of the probandwise rate is similar to that for the pairwise rate, except that the number of doubly ascertained pairs are added to both the numerator and denominator. The most common applications of the probandwise rate are studies in which twins are selected from a clinical population because they are twins. Probandwise rates are useful because they may be compared to estimates of morbid risk commonly used in epidemiology. Although differences in concordance rates between MZ and DZ twins are of interest, when only concordance rates are reported, a good deal of information is generally left unextracted from the data.

Table 7.2 (adapted from Kendler, 1983) contains the concordance rates reported in published twin studies of schizophrenia in which probandwise concordance rates could be determined. (The study by Luxenburger, 1928, is not included because conflicting concordance rates were published and the lack of any concordant DZ twins prevents the exact estimation of genetic influences.) Inspection of the table reveals very different concordance rates across the various studies. Factors such as differences in the severity of the probands' illness, diagnostic practices, and the completeness of ascertainment can all produce such discrepant findings (Kendler, 1983). These factors should not differ, however, between MZ and DZ twins in the same study. This suggests that the most useful basis for comparison among different studies would be a variable derived from MZs and DZs within the same study—G, the coefficient of genetic determination, is such a variable.

In 1965, Falconer described a multifactorial polygenic (MFP) theory of disease transmission that was applied to schizophrenia by Gottesman and Shields in 1967. This model assumes that there exists a normally distributed liability to the development of a disease that reflects the additive effects of numerous genetic and environmental factors. When an individual's liability exceeds a specific threshold on the distribution, he or she manifests the

TABLE 7.2.
Probandwise Concordance and Broad Heritability (G) for Schizophrenia in Twin Studies in which Probandwise Concordance Could Be Exactly Determined

	Proband concordance						Broad estimated heritability (G)
	Monozygotic			Same-sex dizygotic			
		Concordance			Concordance		
Study	N^a	N^b	%	N^a	N^b	%	
Essen-Moller (1941)	11	7	64	27	4	15	.87
Slater (1953)	41	32	78	61	14	23	.65
Inouye (1963)	55	33	60	11	2	18	.66
Kringlen (1967)	69	31	45	96	14	15	.61
Fischer et al. (1969)	23	14	61	43	12	28	.41
Gottesman & Shields (1972)	26	15	58	34	4	12	.86
Allen et al. (1972)	111	42	38	130	11	8	.80
Tienari (1975)	21	7	33	42	6	14	.53
Kendler & Robinette (1983)	194	60	31	277	18	7	.91
Mean (unweighted)							.70

SOURCE: Adapted from Kendler (1983)
[a]Total number of co-twins of primarily ascertained index twins
[b]Number of co-twins of primarily ascertained index twins with schizophrenia

disease. The G statistic estimates the "broad heritability" or the percentage of variance in liability due to the assumed additive genetic factors. The value of G is determined by doubling the difference in correlations between MZ and DZ pairs. For determining twin correlations for a categorical trait, it is necessary to know the prevalence of the trait in the population. Table 7.2 contains the values of G calculated for the included studies. Inspection of these values reveals that the findings are far less discrepant than they appear as raw concordance rates. In fact, Kendler (1983) demonstrated that there are no statistically significant differences between the values of G in any of the nine studies. The unweighted mean of G for all nine studies is .70, which indicates that 70% of the variance in the liability to schizophrenia is due to additive genetic influences. The plausibility of the assumption of an MFP model, upon which the calculation of G is based, will be discussed below.

Twins have also been used for purposes other than estimating the proportion of variance contributed by genetic factors. Twins have been used to attempt to identify "more" and "less" genetic forms of schizophrenia. Dworkin and Lenzenweger (1984) concluded that negative symptoms may characterize a more genetic form of schizophrenia. Gottesman and Shields (1972) found that using diagnostic criteria for schizophrenia of "medium specificity" maximized the ratio of MZ to DZ concordance rates. McGuffin et al. (1984) found that the Research Diagnostic Criteria (RDC) and the Feighner Criteria produced the highest heritability for schizophrenia in a twin study. Farmer et al. (1987) reported that including schizophrenia with mood incongruent delusions, schizotypal personality, and atypical psychosis produced the maximum difference between MZ and DZ concordance. Fischer (1973) investigated the rates of schizophrenia in the offspring of the schizophrenic and nonschizophrenic members of discordant MZ pairs. Her sample, drawn from the Danish Twin Register, was followed up 17 years later by Gottesman and Bertelsen (1989). There was no significant difference in morbid risk for schizophrenia in the offspring of the schizophrenic twins versus the offspring of their nonschizophrenic co-twins. These, and related findings, are reviewed in Lyons et al. (1989).

The overwhelming conclusion from the extant adoption and twin studies is that, to a great extent, the familial influence on schizophrenia is genetic. However, it is important to recognize that the same studies that provide strong evidence for genetic factors also provide the strongest evidence for the role of the environment in the etiology of schizophrenia. This point is made most compellingly by the rate of discordance among MZ twins. That is, when one MZ twin has schizophrenia, in about half the cases, his or her genetically identical co-twin does not have the disorder. The reasons for discordance must be environmental. These, as yet unidentified, environmental influences could range from prenatal viral infection and obstetrical difficulties to deviant communication patterns in the family and parental overinvolvement and criticism. In general, genetic research has provided stronger, albeit nonspecific, evidence for the role of environmental factors than has environmentally oriented research.

What Is Inherited?

A reasonable starting point to address this question is to determine which phenotypes occur in excess among the biological relatives of schizophrenics. The type of phenotype that has been studied most extensively is psychiatric disorder. Clearly, the MR for schizophrenia is elevated among biological relatives of schizophrenic probands. There are a number of other psychiatric phenotypes that are genetically associated with schizophrenia. Kendler (1985) described two historical approaches to the examination of deviant but nonpsychotic relatives of schizophrenics. The first approach focuses on investigation of deviant but nonpsychotic relatives of schizophrenic patients. Rosanoff (1911) observed that relatives were more often "cranky, stubborn; worries over nothing; religious crank, nervous, queer; restless, has phobias; suspicious." Likewise, Bleuler (1924) noted that relatives of schizophrenic individuals were more likely to be "shut-in and eccentric." In the 1920s, Kretschmer (1921; 1925) described the schizoid personality and in the 1930s, Kallmann (1938) identified "borderline cases" and "schizoid psychopathy" as occurring with increased frequency in the relatives of schizophrenics. Additional studies examined the deviant traits and behaviors of nonpsychotic co-twins of schizophrenic probands (Inouye, 1970; Slater, 1953). These studies likewise described deviancy ranging from "typically autistic" to paranoid and eccentric behaviors in the nonschizophrenic twin. The second approach examines the nonpsychotic variants of schizophrenia that are not necessarily familial. These studies include identification of ambulatory schizophrenia (Zilboorg, 1941), psuedoneurotic schizophrenia (Hoch & Polatin, 1949), "as if" personality (Deutsch, 1942), schizotypal personality (Rado, 1953), and "schizotaxia" (Meehl, 1962) as part of the schizophrenia spectrum.

Historically, the term "schizoid" was used to categorize individuals who displayed nonpsychotic symptoms of schizophrenia, such as cognitive idiosyncracies and social isolation, that were presumed to have some relation to schizophrenia (Siever & Gunderson, 1983). Over time, however, "schizoid" simply came to mean cold, aloof, socially awkward, and physically ill at ease. Sandor Rado (1953) introduced the notion of "schizotypal"—meaning "like schizophrenia"—to reassert a putative genetic connection between schizophrenia and certain individuals with personality styles characterized by impaired ego functions, impoverished affect, and limited interpersonal relations.

Since then, theorists and researchers, beginning with Meehl (1962) and Hoch and Polatin (1949), have sought to describe and define a personality type that shares some characteristics of schizophrenia as stable personality traits and has a common genetic substrate—a "borderline schizophrenia" (Kety et al., 1968). DSM-III consolidated and clarified these efforts by creating the diagnostic category Schizotypal Personality Disorder and distinguishing it from Schizoid Personality Disorder on the one hand, and Borderline Personality Disorder on the other (Gunderson et al., 1983; Spitzer et al., 1979; Torgersen, 1984). Kety et al. (1968) suggested that strictly defined schizophrenia represented one end of a spectrum of disorders of varying severity. Instead of viewing schizophrenia as a dichotomous trait that was either present or absent, the spectrum concept views individuals as having varying degrees of vulnerability to schizophrenia ranging from normal individuals through personality-disordered individuals to the extreme of classical schizophrenia. In one of the early Danish studies, Kety et al. (1968) predicted that "if schizophrenia were to some extent genetically transmitted, there should be a higher prevalence of disorders in the schizophrenia spectrum among the biological relatives" (p. 353). Kety did, indeed, find a concentration of schizophrenia spectrum diagnoses in the biological relatives of schizophrenic adoptees. The numbers were, however, too small to permit analysis of the separate diagnoses.

In a later study with a large number of subjects, Kety et al. (1975) again examined the relatives of schizophrenic probands and confirmed the presence of elevated rates of spectrum disorders.

Rosenthal et al. (1968) examined the occurrence of schizophrenia spectrum disorders in the adopted-away children of schizophrenic parents. His concept of schizophrenia and related disorders included both a "hard spectrum" borderline and chronic schizophrenia and a "soft spectrum" that consisted of acute schizophrenia and "schizophrenic personalities" defined as undifferentiated, inadequate, subparanoid, and schizoid. Rosenthal reported an 18.8% prevalence of hard spectrum disorders in the index adoptees compared with 10.1% in the controls; a difference that just misses statistical significance.

Although the early adoption studies supported the notion of a genetically based schizophrenia spectrum, more specific analysis of the disorders that comprise this spectrum was often difficult because of small numbers and/or ill-defined diagnostic criteria. More precise information regarding the boundaries of the schizophrenia spectrum has become available following reanalysis of these early studies and the addition of new ones. In the early studies, investigators generally examined chronic, acute, latent, borderline, and uncertain schizophrenia, as well as schizoid and inadequate personality disorders as candidates for the spectrum. More recent studies have investigated DSM-III defined schizophrenia, schizophreniform disorder, and anxiety and delusional disorders, as well as borderline, paranoid, schizotypal and schizoid personality disorders, as possible components of the schizophrenia spectrum.

There is little dispute that chronic schizophrenia can be considered a part of the spectrum of disorders. The risk for acute schizophrenia, or schizophreniform disorder, does not seem to be elevated in the relatives of schizophrenic probands and it is generally excluded from the spectrum (Ingraham & Kety 1988; Kety et al., 1968). Likewise, there is no empirical basis for including anxiety and delusional disorders in the schizophrenia spectrum (Kendler et al., 1981a; 1981b). The boundaries of the spectrum in regard to personality disorders remain, however, less firmly established.

Kendler and Gruenberg (1984) reanalyzed the Danish adoption studies and found a 16% prevalence of schizotypal and paranoid personality disorders in the biological relatives of schizophrenic adoptees, a rate that is significantly greater than that of the controls. After blindly rediagnosing the relatives of the schizophrenic adoptees according to DSM-III criteria, Kendler et al. (1981c) found schizotypal personality in 10.5% of the biological relatives of schizophrenic probands compared with 1.3% in adoptive relatives and biological relatives of controls. He further reported 3.8% prevalence of paranoid personality disorder in the biologial relatives of index compared with 0.9% in the control cases. These highly significant findings led to the conclusion that paranoid and schizotypal disorders can be included in the spectrum. Similarly, Lowing et al. (1983) reanalyzed the data and found that schizotypal as well as schizoid personalities occurred in significant excess in the adopted-away offspring of schizophrenic probands. Kety et al. (1975), however, reported that schizoid and inadequate personality disorders were equally present in the relatives of schizophrenic and control probands and therefore excluded these from the spectrum. In independent investigations, both Gunderson et al. (1983) and Lowing et al. (1983) excluded borderline personality disorder from the schizophrenia spectrum, finding no differences in rates of the disorder between the relatives of index and control subjects. It therefore appears that there is a significant consensus based on the adoption studies to exclude acute schizophrenia, anxiety disorders, delusional disorder, and borderline personality from the spectrum and general agreement that chronic schizophrenia and schizotypal personality disorder can be included. Although there is evidence supporting the inclusion of schizoid and paranoid personality disorder, disagreement remains. Other, recent psychiatric genetic studies have also supported the spectrum concept (Baron et al., 1985; Kendler et al., 1984; Kendler & Gruenberg, 1985). Ingraham and Kety (1988) and Kendler (1987) have recently reviewed the relevant research. The existence of a genetic spectrum is consistent with both multifactorial polygenic models that posit a continuum of vulnerability and with major locus models and the concept of pleiotropy (a given gene may produce more than one kind of phenotype) or some environmental influence.

In addition to psychiatric disorders, there is evidence that a schizophrenia genotype may lead to other phenotypic outcomes. Holzman and Matthysee (1990) reviewed several of the most promising of these nonpsychiatric phenotypes. Eye movement dysfunction (EMD) has been demonstrated to have an intriguing association with schizophrenia. Eye movement is assessed while a subject visually tracks a target presented on a screen. The accurate tracking of the target depends on the smooth pursuit visual system; in some subjects, tracking is disrupted by intrusions from the saccadic or rapid eye movement system.

Holzman (1987; Holzman et al., 1988) reviewed a substantial body of evidence that supports the use of EMD as a biological marker for schizophrenia. Between 51% and 85% of schizophrenic patients have been found to have EMDs compared with 8% in a normal population. Inattention, poor motivation, and neuroleptic drug treatment have been ruled out as the source of EMD. A high proportion (30% to 50%) of patients with affective disorder have also been found to have EMDs. However, the finding that approximately 45% of the nonschizophrenic first-degree relatives of schizophrenics have EMDs compared with 10% of first-degree relatives of affective patients suggests that it may represent a trait related to schizophrenia and a state related to affective disorder. This is reinforced by the finding of high rates of EMDs among the first-degree relatives of schizophrenic probands who themselves do not show EMD.

Holzman and Levy (1977) found that EMDs tend to be present in at least one clinically unaffected parent of schizophrenic probands with EMD. Among twins discordant for schizophrenia, more than 80% of MZ pairs were concordant for EMD and more than 40% of DZ pairs were concordant for EMD. In some pairs of DZ twins discordant for schizophrenia, the schizophrenic twin had normal smooth pursuit, while the clinically unaffected co-twin had eye tracking abnormalities. Such findings indicate that EMD is not a byproduct of clinical schizophrenia.

Measures of attention and information processing have also been suggested as alternative expressions of a schizophrenia genotype (Holzman & Matthysse, in press). Performance on the continuous performance test (CPT) has been found to be impaired in schizophrenics. Versions of the CPT that present a more difficult task—through the use of multiple or degraded stimuli—than the conventional CPT have demonstrated impaired performance among groups at elevated risk for schizophrenia, such as the offspring of schizophrenic parents (Erlenmeyer-Kimling & Cornblatt, 1978; Nuechterlein, 1983). Impaired CPT performance has also been found in nonclinical samples identified by psychometric measures of putative vulnerability to schizophrenia. Another attentional measure, the span of apprehension technique, has produced similar findings with regard to impairment observed in schizophrenics, their unaffected relatives, and subjects defined psychometrically as being at elevated risk for schizophrenia. Abnormalities of cortical evoked potentials have also been suggested as possibly related to the schizophrenia genotype (Friedman et al., 1982; Josiessen et al., 1985; Saitoh et al., 1984).

High Risk Studies

High risk studies, which involve the study of individuals believed to be at elevated risk for developing schizophrenia, are another approach to answering the question "what is inherited?" High risk populations for prospective study frequently consist of the offspring of schizophrenic parents whose morbid risk is estimated at 12.8% for children with a single affected parent and 46.3% when both parents are schizophrenic (Reich, 1976). Retrospective reports of the childhood behavior of adults with schizophrenia spectrum disorders have also been used in high risk studies.

Characteristics identified in high risk populations may be viewed as vulnerability indicators that "reflect a familially transmitted alteration in neurological process that constitutes a vulnerability or diathesis to subsequent schizophrenia" (Heinrichs & Buchanan, 1988:16). These childhood neurological signs may or may not be evident in adults with schizophrenia spectrum disorders and are not necessarily related to the thought and behavior disturbances associated with the pathology. Childhood psychosis is not considered a precursor to schizophrenia since "a meaningful distinction can be made between psychosis that begins in the early years of life and psychosis that begins at or after puberty" (Gottesman & Shields, 1982:163). Childhood schizophrenia and infantile autism do not appear to be genetically related to schizophrenia and are therefore considered neither a part of the spectrum of related disorders nor a means of identifying a high risk population useful in the study of adult schizophrenia.

Some of the proposed vulnerability markers observed in high risk populations include deficits in social behavior, formal cognition, vigilance and attention, and neurointegrative behaviors. In his reanalysis of the Danish adoption studies, Kendler et al. (1982)

examined the relationship between childhood behavioral disturbance and adult schizophrenia spectrum disorders. He found that childhood social withdrawal and antisocial behavior were more commonly reported in the retrospective histories of biological relatives of schizophrenic individuals. Kendler concluded that childhood withdrawal is genetically linked to both adult schizophrenia and schizotypal personality disorder.

In longitudinal studies of children of schizophrenic mothers, Mednick and Schulsinger (1965) and Schulsinger (1976) reported that children who developed either schizophrenia or schizotypal personality disorder had premorbid marker behaviors that included defective emotional rapport and formally disturbed cognition. They further observed that those who developed schizophrenia had more severe attention deficits and poorer affective control than those who developed schizotypal personalities. The authors concluded that schizophrenia and schizotypal personality disorders are genetically related. Weintraub (1987) likewise followed the children of one or two schizophrenic parents longitudinally and found that attention and information-processing deficits were the most sensitive predictors of subsequent pathology; he further noted that these behaviors differentiated children who were to develop schizophrenia and schizotypal personality disorders from those who would not.

Neurological abnormalities such as vigilance deficits have been suggested as vulnerability indicators in the children of schizophrenic and schizotypal adults. Nuechterlein (1983) studied results of a continuous performance task and found that these children had more errors than normals, particularly those involving perceptual sensitivity. Fish (1987) followed offspring of schizophrenic individuals from infancy and discovered a pattern of "erratic neurointegrative development over time" that was predictive of subsequent pathology. The severity of this "pandysmaturation" was related to psychopathology at 10 years and was predictive of schizophrenia and schizotypal and paranoid personality disorders. Numerous investigators have likewise considered fine motor deficits, right-left orientation, impaired motor coordination, sensory overflow, hypoactivity, and sensory integration to be neurological signs in the relatives of schizophrenics and in those who subsequently develop schizophrenia (Kinney et al., 1986; Marcus, 1974; Marcus et al., 1985; Rieder & Nichols, 1979).

These neurological signs appear to represent a diathesis to schizophrenia and contribute to the delineation of the boundaries of the spectrum. As previously mentioned, studies of the children of schizotypal and schizophrenic adults support the inclusion of schizotypal personality disorder in the spectrum (Fish, 1987; Nuechterlein, 1983; Schulsinger, 1976). In her prospective study of neurointegrative deficits in a high risk population, Fish (1987) presented data that further suggest that paranoid personality disorder may also be genetically related to the schizophrenic disorders.

How Is It Inherited?

The third critical question about the genetics of schizophrenia is to determine the mode (or modes) of genetic transmission. A number of models have been proposed to describe the genetic mode of transmission of schizophrenia and related disorders. Early investigators, such as Rudin (1916) and Kallmann (1938), suggested that simple Mendelian genetics accounted for the transmission of the disorder; this model proposed that a single major locus, whether a dominant or two recessive alleles, accounted for the genetic contribution to the development of schizophrenia. Since then, there have been numerous approaches to describing the mode of inheritance. The two major approaches are modifications of the single major locus model (SML) and the multifactorial polygenic model (MFP). However, most modern investigators have concluded that clinically defined schizophrenia does not seem to follow a neat Mendelian pattern. Therefore, a number of modifications to the classical Mendelian model have been proposed. The first is reduced penetrance. Penetrance, a statistical, rather than biological, construct, is a probabilistic statement about the proportion of individuals with a certain genotype who manifest that genotype in their phenotypes. (This should not be confused with expressivity, which describes the degree to which an individual manifests a genotype in his or her phenotype.) Reduced penetrance describes a phenomenon in which not all individuals with a certain genotype will express it phenotypically. The concept of phenocopies can also explain observed departures from Mendelian predictions about transmission. A phenocopy is a condition that mimics or resembles a phenotype produced by specific genetic factors. For example, Wender and Klein (1981) pointed out that the effect on fetal limb development of prenatal exposure to the drug thalidomide (phocomelia) is a phenocopy of a rare genetic disorder, acheiropody. This provides a complement to the notion of reduced penetrance; that is, although genotype X generally produces phenotype Y, not all individuals with genotype X will manifest phenotype Y (reduced penetrance) and not all individuals with phenotype Y will have genotype X (phenocopy).

In general, SML models have not fared well in statistical tests of their adequacy for describing empirical results. Following a review of mathematical models of genetic transmission, Faraone et al. (1988) concluded that the results of prevalence and segregation analyses do not support the SML model. Although many studies of the model did not include statistical tests of its adequacy, those that did rejected the SML model.

O'Rourke et al. (1982) and McGue et al. (1985) have reported analyses that they used to rule out a single gene model. However, Diehl and Kendler (1989) have criticized this conclusion on a number of issues: (1) the early family data that they utilized are not based on clear diagnostic criteria and blind diagnosis; (2) their pooled samples may encompass substantial heterogeneity; (3) results are not corrected for fitness effects; and (4) genetic homogeneity is required for all cases of schizophrenia. Garver et al. (1989) suggested that the "simple" SML model that assumes "that *all* cases of schizophrenia have a common etiology and that *no* familial resemblance is environmentally determined" (p. 424) should be rejected.

Kendler (1988b) identified a number of factors that may contribute to the difficulty in adducing clear-cut evidence for the SML model. The existence of genetic heterogeneity (i.e., more than one genetic type of schizophrenia) would greatly complicate the identification of a single gene. Environmental influences that could produce phenocopies or reduce penetrance would make it difficult to demonstrate the presence of a single gene influence. The action of a single gene would be difficult to recognize if the familial environment also contributed to family resemblance. The presence of genotype X environment interaction, in which the gene influences the impact of the environment, would complicate analyses to detect the influence of a single gene. There are also a number of technical issues, such as the misclassification of phenotypes, differential mortality and migration, and reduced reproductive fitness, that substantially complicate efforts to identify the action of a single gene.

The MFP model, as described above in the context of twin studies, assumes that schizophrenia occurs when an individual's liability for schizophrenia exceeds a certain threshold. The liability is assumed to be normally distributed in the population and to reflect the additive effect of numerous genetic and environmental factors. A number of points that support the application of an MFP model to schizophrenia have been suggested (Gottesman & Shields, 1967; Hanson et al., 1977): (1) schizophrenia, like MFP disorders, occurs with varying degrees of severity; (2) the risk of schizophrenia is correlated with the number of affected relatives; (3) the existence of nonschizophrenic spectrum disorders is compatible with individuals with high, but subthreshold, liability; and (4) natural selection has a slower influence on MFP disorders.

There have been a number of impressive successes fitting the MFP model to data, especially using path analysis (McGue et al., 1983; 1985; McGue et al., 1987; Rao et al., 1981). There have also been some failures, with several studies rejecting versions of the MFP model (Baron, 1982; Cloninger, 1989; Tsuang et al., 1983).

Linkage Studies

Although much of the research addressing the issue of how schizophrenia is inherited has focused on phenotypes and endophenotypes, with recent advances in molecular genetics it is possible to include the actual genotype in such investigations. The application of linkage studies in schizophrenia has been reinvigorated by the exciting developments in molecular biology (see the chapters by O'Donovan & McGuffin and Risch). The application of restriction fragment length polymorphisms (RFLPs) increases immeasurably the loci that may be evaluated for linkage to schizophrenia. The RFLP approach uses restriction enzymes to cleave DNA into fragments of different lengths. This provides an extremely large number of polymorphic loci that may potentially serve as markers for a gene or genes related to schizophrenia.

There were some linkage studies of schizophrenia conducted before the advent of molecular approaches. As described by O'Donovan and McGuffin in this volume, a number of these studies investigated linkage using HLA markers. Following an initial positive finding (Turner, 1979), there were a number of failures to replicate (Andrew et al., 1987; Chadda et al., 1986; Goldin et al., 1987; McGuffin et al., 1983). O'Donovan and McGuffin conclude that linkage studies using classical, nonmolecular approaches have not been very informative. They suggest that association studies indicate the possibility of a significant, albeit modest, association between schizophrenia and HLA-A9.

The most recent era in linkage studies was initiated in 1988, when Bassett et al. reported a case study of a family in which a partial trisomy of chromosome 5, mild facial dysmorphology, and schizophrenia cosegregated in two members. When the proband was admitted to the hospital with DSM-III-R schizophrenia, the facial abnormalities were noted. Later, the proband's mother commented that her brother had schizophrenia and physically resembled her son, the proband. This motivated a cytogenetic investigation of the family that revealed that both members with schizophrenia also had an unbalanced translocation of material from the 5q11.2 to the 5q13.3 segment of chromosome 5. The proband's mother had a balanced translocation (i.e., no trisomy) of the same region unlike the schizophrenic members of the family, suggesting that the third copy of the chromosome 5 section was responsible for the physical abnormalities and the schizophrenia. It is interesting to note that the two schizophrenic members of the family had abnormalities of smooth pursuit eye movement (SPEM), whereas three unaffected members, including the mother with a balanced translocation, had normal SPEM (Iacono et al., 1988).

The report by Bassett et al. (1988) was followed by a report by Sherrington et al. (1988). These investigators studied five Icelandic and two English families with high concentrations of schizophrenic individuals. The entire sample comprised 104 individuals; 39 had schizophrenia or schizophrenia spectrum disorders and 5 had schizoid personality. These investigators decided to study only medium to large pedigrees with cases of schizophrenia in at least three generations permitting the determination of lod scores within individual families. The families were selected for a high density of cases of schizophrenia. They excluded families with bipolar disorder and selected families with only one possible source of the schizophrenia gene. Sherrington et al. (1988) reported linkage of two loci on chromosome 5 (D5S39 and D5S76) to schizophrenia. These investigators found that, as they expanded the definition of the affected phenotype from traditional schizophrenia to include fringe phenotypes, the odds for linkage, as reflected in the lod score, increased. Gurling et al. (1989) pointed out that, while their "fringe phenotypes" included cases that appeared to be depression or other neurotic disorder, there was clear evidence from the clinical evaluation for the presence of schizophrenia-related psychological abnormalities not covered by the RDC (Spitzer et al., 1978).

There have been two linkage studies reported that have failed to replicate the finding of linkage to chromosome 5. Kennedy et al. (1988) studied a single, large Swedish pedigree with 31 schizophrenic individuals. They failed to find linkage to any of seven loci in the region of chromosome 5 studied by Sherrington. Kaufmann et al. (1989) studied four large North American families that each had two siblings with chronic schizophrenia. There was a total of 17 individuals with schizophrenia in the entire sample as well as a number of schizophrenia spectrum disorders. These investigators were able to rule out a susceptibility locus for schizophrenia on chromosome 5q.

The reason for the failure by Kennedy et al. (1988) and Kaufmann et al. (1989) to replicate the finding of Sherrington et al. (1988) is uncertain. One possibility is that schizophrenia is genetically heterogeneous. Some cases may be related to a gene on chromosome 5, and other cases related to some other locus, a polygenic process, or an environmental etiology. Another possibility that must always be considered when there is a failure to replicate a finding is that the original finding may be a Type I error. Given the growing number of failures to replicate Sherrington et al.'s original finding, the latter possibility must be seriously considered.

Segregation Studies

The mode of transmission for schizophrenia has been investigated through complex segregation analysis. Diehl and Kendler (1989) and Vogler et al. (1990) have reviewed the published segregation analyses of schizophrenia. Tsuang et al. (1982) and Elston et al. (1978) found evidence for vertical transmission but were able to reject single gene transmission (but Risch and Baron (1984) identified methodological problems that would qualify these conclusions).

Carter and Chung (1980) and Risch and Baron (1984) found that both polygenes with relatively high heritability or a single major locus model were compatible with their results. Rao et al. (1981) used data pooled from numerous family studies that met strict inclusion criteria and found the observed pattern of inheritance consistent with a multifactorial threshold model. Vogler et al. (1990) reported the results of a mixed model segregation analysis of schizophrenia using a Swedish sample reported by Lindelius (1970). They found three major gene models to be less acceptable than a multifactorial model without major gene effects. Numerical instabilities prevented a stable solution to the basic mixed model. In all of the models that allowed for a major gene, the additive model was preferred over recessive or dominant major gene models. Numerical instabilities precluded formal hypothesis testing. Diehl and Kendler (1989) concluded that the power of segregation analysis to resolve the transmission of a "complex" human genetic disorder is unclear (Reich et al., 1981).

Conclusions

One of the major unanswered questions about the genetics of schizophrenia is the mode of transmission. The primary alternatives are the MFP model and some version of the SML model. In general, the MFP model has been more successful at accommodating empirical findings. Some have suggested that this may be due to the fact that MFP models are more inherently "accommodating." For example, Medawar (1967) suggested that multifactorial theories are like "analgesic pills which dull the aches of incomprehension without going to their causes" (as quoted in Dalén, 1972, p. 481). However, the mere fact that a theory is flexible is not a substantive basis for criticizing it. Also, modern statistical techniques, such as path analysis, allow for rigorous testing of MFP models. Another nonsubstantive criticism of the MFP model is that it represents a "biological null hypothesis." This refers to the probable difficulty in associating a specific pathophysiological process with a multifactorial etiology. If a single gene responsible for schizophrenia could be identified, it might then be rather straightforward to identify the gene product and determine how it produces the clinical phenotype. An important reason for pursuing linkage studies aggressively is the promise that they hold for providing information that would make schizophrenia amenable to more effective treatments and prevention. Gottesman and McGue (1990), however, have suggested that MFP models may not be as pessimistic as they are sometimes held to be. They suggest that cardiovascular disease (CVD), which seems to be due to an MFP process, may be a useful analogue. In CVD research, the identification of pathophysiological mechanisms influencing lipids has provided the opportunity to detect associations between specific genetic loci and the clinical outcome. They suggest that improved characterization of the pathophysiologies underlying schizophrenia will provide phenotypes more amenable to molecular genetic analysis. Gottesman and Shields (1982) have also pointed out that "whenever polygenic variation has been studied under *laboratory conditions* . . . a few handleable genes have proved to mediate a large part of the genetic variance" (pp. 225–226).

Proponents of the MFP model are not alone in using analogies to other medical disorders to support the plausibility of their view. Gurling et al. (1989) make the point that a number of individuals have suggested that schizophrenia is not amenable to linkage techniques that have been used to detect major genes for Mendelian disorders because the evidence for single gene transmission is untenable (McGuffin & Sturt, 1986; O'Rourke et al., 1982; Sturt & McGuffin, 1985). Gurling et al. point out that similar objections were never raised about medical disorders such as multiple endocrine neoplasia (MEN2a), which also demonstrates incomplete penetrance and "apparent polygenic or uncertain inheritance." Clearly, the advent of molecular genetic techniques has given great impetus to the investigation of SML models. Also, approaches such as the very promising latent trait model described by Holzman and Matthysse have brought new vigor to the pursuit of evidence to support an SML model. (See the chapter by Matthysse for more information on this approach.)

In their review of mathematical models of the genetic transmission of schizophrenia, Faraone et al. (1988) concluded that "given the powerful analytic methods already available, it is unlikely that advances in statistical analysis will resolve the problems of genetic heterogeneity" (p. 527). In a similar vein, Holzman and Matthysse (1990) asserted that recurrence risk and segregation analysis have taken us as far as they are capable of taking us.

Several investigators in the field have suggested that sole

reliance on the clinical phenotypes associated with schizophrenia is unlikely to lead to further progress in our genetic understanding of its transmission. Holzman and Matthysse (1990) have commented that the numerous revisions and refinements in the definition of schizophrenia amount to reshuffling and redealing the "same deck of symptom cards." They advocate the inclusion of nonpsychiatric phenotypes in the genetic study of schizophrenia, such as their own work using abnormalities of smooth pursuit eye movement in the context of a latent trait model. There are a number of reasons why better specification of phenotypes (and endophenotypes) associated with a schizophrenia genotype(s) could be very productive: (1) the specification of pathophysiological and/or psychological "markers" could help to resolve genetic heterogeneity in schizophrenia (if it exists) and thereby promote progress in molecular genetic approaches; (2) an improved specification of nonclinical phenotypes associated with a schizophrenia genotype would greatly improve the classification of "affected" individuals for linkage studies; and (3) the explication of pathophysiological processes underlying schizophrenia could facilitate the investigation of the influence of polygenic factors on the occurrence of schizophrenia.

References

Abrams, R. & Taylor, M. A. (1983) The genetics of schizophrenia: A reassessment using modern criteria. *American Journal of Psychiatry* 140:171–175.

Allen, M. G., Cohen, S. & Pollin, W. (1972) Schizophrenia in veteran twins: A diagnostic review. *American Journal of Psychiatry* 128:939–945.

Allen, M. G., & Pollin, W. (1970) Schizophrenia in twins and the diffuse ego boundary hypothesis. *American Journal of Psychiatry* 127:437–442.

Andreasen, N. C., Endicott, J., Spitzer, R. L. & Winokur, G. (1977) The family history method using diagnostic criteria: Reliability and validity. *Archives of General Psychiatry* 34:1229–1235.

Andrew, B., Watt, D. C., Gillespie, C. & Chapel, H. (1987) A study of genetic linkage in schizophrenia. *Psychological Medicine* 17:363–370.

Baron, M. (1982) Genetic models of schizophrenia. *Acta Psychiatrica Scandinavica* 65:263–275.

Baron, M., Gruen, R., Rainer, J. D., Kane, J., Asnis, L. & Lord, S. (1985) A family study of schizophrenic and normal control probands: Implications for the spectrum concept of schizophrenia. *Biological Psychiatry* 20:94–119.

Bassett, A. S., McGillivray, B. C., Jones, B. D. & Pantzar, J. T. (1988) Partial trisomy chromosome 5 cosegregation with schizophrenia. *Lancet* i:799–800.

Bleuler, E. (1924) *Textbook of Psychiatry* (A. A. Brill Trans.). New York: Macmillan.

Carter, C. L. & Chung, C. S. (1980) Segregation analysis of schizophrenia under a mixed genetic model. *Human Heredity* 30:350–356.

Chadda, R., Kullhara, P., Singh, T. & Sehgal, P. S. (1986) HLA antigens in schizophrenia: A family study. *British Journal of Psychiatry* 149:612–615.

Cloninger, C. R. (1989) Schizophrenia: Genetic etiological factors. In *Comprehensive Textbook of Psychiatry*, ed. H. I. Kaplan & B. J. Sadock. Baltimore: Williams & Wilkins.

Coryell, W. & Zimmerman, M. (1988) The heritability of schizophrenia and schizoaffective disorder. *Archives of General Psychiatry* 45:323–327.

Dalén, P. (1972) One, two, or many? In *Genetic Factors in Schizophrenia*, ed. A. R. Kaplan. Springfield, IL: Charles C. Thomas.

Deutsch, H. (1942) Some forms of emotional disturbance and their relationship to schizophrenia. *Psychoanalytic Quarterly* 11:301–321.

Diehl, S. C. & Kendler, K. S. (1989) Strategies for linkage studies of schizophrenia: Pedigrees, DNA markers, and statistical analyses. *Schizophrenia Bulletin* 1:403–419.

Dworkin, R. H. & Lenzenweger, M. F. (1984) Symptoms and genetics of schizophrenia: Implications for diagnosis. *American Journal of Psychiatry* 141:1541–1545.

Elston, R. C., Namboodiri, K. K., Spence, M. A. & Rainer, J. D. (1978) A genetic study of schizophrenia pedigrees. One-locus hypotheses. *Neuropsychobiology* 4:193–206.

Erlenmeyer-Kimling, L. & Cornblatt, B. (1978) Attentional measures in a study of children at high risk for schizophrenia. In *The Nature of Schizophrenia: New Approaches to Research and Treatment*, ed. L. C. Wynne, R. L. Cromwell & S. Matthysse. New York: John Wiley.

Essen-Moller, E. (1941) Psychiatrische Untersuchungen an einer Serie von Zwillingen. *Acta Psychiatrica Scandinavica* 23 (Suppl.).

Falconer, D. S. (1965) The inheritance of liability to certain diseases estimated from the incidence among relatives. *Annals of Human Genetics* 29:51–76.

Faraone, S. V., Lyons, M. J. & Tsuang, M. T. (1988) Mathematical models of genetic transmission. In *Handbook of Schizophrenia*, Vol. 3: *Nosology, Epidemiology, and Genetics*, ed. M. T. Tsuang & J. C. Simpson) Amsterdam: Elsevier.

Farmer, A. E., Jackson, R., McGuffin, P. & Storey, P. (1987) Cerebral ventricular enlargement in chronic schizophrenia. *British Journal of Psychiatry* 150:324–330.

Fischer, M. (1973) Genetic and environmental factors in schizophrenia. *Acta Psychiatrica Scandinavica* 238 (Suppl.).

Fischer, M., Harvald, B. & Hauge, M. (1969) A Danish twin study of schizophrenia. *British Journal of Psychiatry* 115:981–990.

Fish, B. (1987) Infant predictors of the longitudinal course of schizophrenic development. *Schizophrenia Bulletin* 13:395–409.

Frangos, E., Athanassenas, G., Tsitourides, S., Katsanov, N. & Alexandrakov, P. (1985) Prevalence of DSM-III schizophrenia among the first-degree relatives of schizophrenic probands. *Acta Psychiatrica Scandinavica* 72:382–386.

Friedman, D., Vaughan, H. G. & Erlenmeyer-Kimling, L. (1982) Cognitive brain potentials in children at risk for schizophrenia. *Schizophrenia Bulletin* 8:514–531.

Garver, D. L., Reich, T., Isenberg, K. E. & Cloninger, C. R. (1989) Schizophrenia and the question of genetic heterogeneity. *Schizophrenia Bulletin* 15:421–430.

Gershon, E. S., DeLisi, L. E., Hamovit, J., Nurnberger, J. I., Jr., Maxwell, E., Schreiber, J., et al. (1988) A controlled family study of chronic psychoses. *Archives of General Psychiatry* 45:328–336.

Goldin, L. R., DeLisi, L. F. & Gershon, E. S. (1987) The relationship of HLA to schizophrenia in 10 nuclear families. *Psychiatry Research* 20:69–78.

Gottesman, I. I. (1978) Schizophrenia and genetics: Where are we? Are you sure? In *The Nature of Schizophrenia: New Approaches to Research and Treatment*, ed. L. C. Wynne, R. L. Cromwell & S. Matthysse. New York: John Wiley.

Gottesman, I. I. & Bertelsen, A. (1989) Confirming unexpressed genotypes for schizophrenia: Risks in the offspring of Fischer's Danish identical and fraternal discordant twins. *Archives of General Psychiatry* 46:867–872.

Gottesman, I. I. & McGue, M. (1990) Mixed and mixed-up models for the transmission of schizophrenia. In *Thinking Clearly About Psychology: Essays in Honor of Paul E. Meehl*, ed. D. Cichetti & W. Grove. Minneapolis: University of Minnesota Press.

Gottesman, I. I. & Shields, J. (1967) A polygenic theory of schizophrenia. *Proceedings of the National Academy of Sciences* 58:199–205.

Gottesman, I. I. & Shields, J. (1972) *Schizophrenia and Genetics: A Twin Study Vantage Point*. New York: Academic Press.

Gottesman, I. I. & Shields, J. (1982) *Schizophrenia: The Epigenetic Puzzle*. New York: Cambridge University Press.

Gunderson, J. G., Siever, L. J. & Spaulding E. (1983) The search for a schizotype: Crossing the border again. *Archives of General Psychiatry* 40:15–22.

Gurling, H. M. D., Sherrington, R. P., Brynjolfsson, J., Read, T., Curtis, D., Mankoo, B. J., et al. (1989) Recent and future molecular genetic research into schizophrenia. *Schizophrenia Bulletin* 15:373–382.

Guze, S. B., Cloninger, C. R., Martin, R. L. & Clayton, P. J. (1983) A follow-up and family study of schizophrenia. *Archives of General Psychiatry* 40:1273–1276.

Hanson, D. R., Gottesman, I. I. & Meehl, P. (1977) Genetic theories and the validation of psychiatric diagnoses: Implications for the study of children of schizophrenics. *Journal of Abnormal Psychology* 86:575–588.

Heinrichs, D. W. & Buchanan, R. W. (1988) Significance and meaning of neurological signs in schizophrenia. *American Journal of Psychiatry* 145:11–18.

Heston, L. L. (1966) Psychiatric disorders in foster home reared children of schizophrenic mothers. *British Journal of Psychiatry* 112:819–825.

Hoch, P. H. & Polatin, P. (1949) Pseudoneurotic forms of schizophrenia. *Psychiatric Quarterly* 23:248–276.

Holzman, P. S. (1987) Recent studies of psychophysiology in schizophrenia. *Schizophrenia Bulletin* 13:49–75.

Holzman, P. S., Kringlen, E., Matthysse, S., Flanagan, S. D., Lipton, R. B., Cramer, G., et al. (1988) A single dominant gene can account for eye tracking dysfunctions and schizophrenia in offspring of discordant twins. *Archives of General Psychiatry* 45:641–647.

Holzman, P. S. & Levy, D. L. (1977) Smooth-pursuit eye movements and functional psychoses: A review. *Schizophrenia Bulletin* 3:15–27.

Holzman, P. S. & Matthysse, S. (1990) Review: The genetics of schizophrenia. *Psychological Science* 1:279–286.

Iacono, W. G., Bassett, A. S. & Jones, B. D. (1988) Eye tracking dysfunction is associated with partial trisomy chromosome 5 and schizophrenia. *Archives of General Psychiatry* 45:1140–1141.

Ingraham, L. J. & Kety, S. S. (1988) Schizophrenia spectrum disorders. In *Handbook of Schizophrenia*, Vol. 3: *Nosology, Epidemiology, and Genetics*, ed. M. T. Tsuang & J. C. Simpson. Amsterdam: Elsevier.

Inouye, E. (1963) Similarity and dissimilarity of schizophrenia in twins. In *Proceedings, Third World Congress of Psychiatry*, vol. 1. Montreal: University of Toronto Press.

Inouye, E. (1970) Personality deviation seen in monozygotic co-twins of the index cases with classical schizophrenia. *Acta Psychiatrica Scandinavica* 219 (Suppl.):90–96.

Josiessen, R. C., Shagass, C., Roemer, R. A. & Straumanis, J. J. (1985) Attention-related effects of somatosensory evoked potentials in college students at high risk for psychopathology. *Journal of Abnormal Psychology* 94:507–518.

Kallmann, F. J. (1938) *The genetics of schizophrenia*. New York: Augustin.

Kaufmann, C. A., DeLisi, L. E., Lehner, T. & Gilliam, T. C. (1989) Physical mapping, linkage analysis of a putative schizophrenia locus on chromosome 5q. *Schizophrenia Bulletin* 15:441–452.

Kendler, K. S. (1983) Overview: A current perspective on twin studies of schizophrenia. *American Journal of Psychiatry* 140:1413–1425.

Kendler, K. S. (1985) Diagnostic approaches to schizotypal personality disorder: A historical perspective. *Schizophrenia Bulletin* 11:538–553.

Kendler, K. S. (1987) The genetics of schizophrenia: A current perspective. In *Psychopharmacology: The Third Generation of Progress*, ed. H. Meltzer, et al. New York: Raven Press.

Kendler, K. S. (1988a) Familial aggregation of schizophrenia and schizophrenia spectrum disorders. *Archives of General Psychiatry* 45:377–383.

Kendler, K. S. (1988b) The genetics of schizophrenia: An overview. In *Handbook of Schizophrenia*, Vol. 3: *Nosology, Epidemiology, and Genetics*, ed. M. T. Tsuang, & J. C. Simpson. New York: Elsevier.

Kendler, K. S. (1988c) Indirect vertical cultural transmission: A model for environmental parental influence on the liability to psychiatric illness. *American Journal of Psychiatry* 145:657–665.

Kendler, K. S. & Gruenberg, A. M. (1984) An independent analysis of the Danish adoption study of schizophrenia. *Archives of General Psychiatry* 41:555–564.

Kendler, K. S., Gruenberg, A. M. & Strauss, J. S. (1981a) An independent analysis of the Copenhagen sample of the Danish adoption study of schizophrenia. I. The relationship between anxiety disorder and schizophrenia. *Archives of General Psychiatry* 38:973–977.

Kendler, K. S., Gruenberg, A. M. & Strauss, J. S. (1981b) An independent analysis of the Copenhagen sample of the Danish adoption study of schizophrenia. III. The relationship between paranoid psychosis (delusional disorder) and schizophrenia. *Archives of General Psychiatry* 38:985–987.

Kendler, K. S., Gruenberg, A. M. & Strauss, J. S. (1981c) An independent analysis of the Copenhagen sample of the Danish adoption study of schizophrenia. II. The relationship between schizotypal personality disorder and schizophrenia. *Archives of General Psychiatry* 38:982–984.

Kendler, K. S., Gruenberg, A. M. & Strauss, J. S. (1982) An independent analysis of the Copenhagen sample of the Danish adoption study of schizophrenia. V. The relationship between childhood social withdrawal and adult schizophrenia. *Archives of General Psychiatry* 39:1257–1261.

Kendler, K. S., Gruenberg, A. M. & Tsuang, M. T. (1985) Psychiatric illness in the first-degree relatives of schizophrenic and surgical control patients: A family study using DSM-III criteria. *Archives of General Psychiatry* 42:770–779.

Kendler, K. S., Masterson, C., Ungaro, R. & Davis, K. (1984) A family history study of schizophrenia-related personality disorder. *American Journal of Psychiatry* 141:424–427.

Kendler, K. S. & Robinette, C. D. (1983) Schizophrenia in the National Academy of Sciences-National Research Council Twin Registry: A 16-year update. *American Journal of Psychiatry* 140:1551–1563.

Kennedy, J. L., Giuffra, L. A., Moises, H. W., Cavalli-Sforza, L. L., Pakstis, A. J., Kidd, J. R., et al. (1988) Evidence against linkage of schizophrenia to markers on chromosome 5 in a northern Swedish pedigree. *Nature* 336:167–170.

Kety, S. S., Rosenthal, D., Wender, P. H. & Schulsinger, F. (1968) The types and prevalence of mental illness in the biological and adoptive families of adopted schizophrenics. *Journal of Psychiatric Research* 1 (Suppl.):345–362.

Kety, S. S., Rosenthal, D., Wender, P. H., Schulsinger, F. & Jacobsen, B. (1975) Mental illness in the biological and adoptive families of adopted individuals who have become schizophrenic: A preliminary report based on psychiatric interviews. In *Genetic Research in Psychiatry*, ed. R. R. Fieve, D. Rosenthal & H. Brill. Baltimore: Johns Hopkins University Press.

Kinney, D. K., Woods, B. T. & Yurgelum-Todd, D. (1986) Neurological abnormalities in schizophrenic patients and their families. *Archives of General Psychiatry* 43:665–668.

Kraepelin, E. (1971) *Dementia Praecox and Paraphrenia* (R. M. Barclay Trans.). Huntington, N.Y.: Krieger. (Original work published 1919)

Kretschmer, E. (1925) *Physique and Character: An Investigation of the Nature of Constitution and of the Theory of Temperament* (2nd ed.) (W. J. H. Sprott Trans.). New York: Harcourt, Brace.

Kretschmer, E. (1970) *Physique and Character: An Investigation of the Nature of Constitution and of the Theory of Temperament* (E. Miller Trans.). New York: Cooper Square Publishers. (Original work published 1921.)

Kringlen, E. (1967) *Heredity and Environment in the Functional Psychoses: An Epidemiological-Clinical Twin Study*. Oslo: Universitetsforlaget.

Lindelius, R. (Ed.) (1970) A study of schizophrenia: A clinical, prognostic, and family investigation. *Acta Psychiatrica Scandinavica* 216 (Suppl.).

Lowing, P. A., Mirsky, A. F. & Pereira, R. (1983) The inheritance of schizophrenia spectrum disorders: A reanalysis of the Danish adoptee study data. *American Journal of Psychiatry* 140:1167–1171.

Luxenburger, H. (1928) Vorlaufiger Bericht uber psychiatrische Serienuntersuchungen an Zwillingen. *Zeitschrift fur die gesamte Neurologie und Psychiatre* 116:297–326.

Lyons, M. J., Kremen, W. S., Tsuang, M. T. & Faraone, S. V. (1989) Investigating putative genetic and environmental forms of schizophrenia: Methods and findings. *International Review of Psychiatry* 1:259–276.

Lytton, H. (1977) Do parents create, or respond to, differences in twins? *Developmental Psychology* 13:456–459.

Marcus, J. (1974) Cerebral functioning in the offspring of schizophrenics: A possible genetic factor. *International Journal of Mental Health* 3:57–73.

Marcus, J., Hans, S. L., Lewow, E., Wilkinson, L. & Burack, C. M. (1985) Neurological findings in high risk children: Childhood assessment and 5-year followup. *Schizophrenia Bulletin* 11:85–100.

Matheny, A. P. (1979) Appraisal of parental bias in twin studies: Ascribed zygosity and IQ differences in twins. *Acta Beneticae Medicae et Gemellologiae* 28:155–160.

McGue, M., Gottesman, I. I. & Rao, D. C. (1983) The transmission of schizophrenia under a multifactorial threshold model. *American Journal of Human Genetics* 35:1161–1178.

McGue, M., Gottesman, I. I. & Rao, D. C. (1985) Resolving genetic models for the transmission of schizophrenia. *Genetic Epidemiology* 2:99–110.

McGue, M., Wette, R. Y. & Rao, D. C. (1987) A Monte Carlo evaluation of three statistical methods used in path analysis. *Genetic Epidemiology* 4:129–155.

McGuffin, P., Farmer, A. I., Gottesman, I. I., Murray, R. M. & Reveley, A. M. (1984) Twin concordance for operationally defined schizophrenia: Confirmation of familiality and heritability. *Archives of General Psychiatry* 41:541–545.

McGuffin, P., Festenstein, H. & Murray, R. M. (1983) A family study of HLA antigens and other genetic markers in schizophrenia. *Psychological Medicine* 13:31–43.

McGuffin, P. & Sturt, E. (1986) Genetic markers in schizophrenia. *Human Heredity* 36:65–88.

Medawar, P. B. (1967) *The Art of the Soluble*. London: Methuen.

Mednick, S. A. & Schulsinger, F. (1965) A longitudinal study of children with a high risk for schizophrenia: A preliminary report. In *Methods and Goals in Human Behavior Genetics*, ed. S. Vandenberg. New York: Academic Press.

Meehl, P. E. (1962) Schizotaxia, schizotypy, schizophrenia. *American Psychologist* 17:827–838.

Nuechterlein, K. H. (1983) Signal detection in vigilance tasks and behavioral attributes among offspring of schizophrenic mothers and among hyperactive children. *Journal of Abnormal Psychology* 92:4–28.

O'Rourke, D. H., Gottesman, I. I., Suarez, B. K., Rice, J. & Reich, T. (1982) Refutation of the general single-locus model for the etiology of schizophrenia. *American Journal of Human Genetics* 34:630–649.

Pope, H. G., Jones, J. M., Cohen, B. M. & Lipinski, J. F. (1982) Failure to find evidence of schizophrenia in first-degree relatives of schizophrenic probands. *American Journal of Psychiatry* 139:826–830.

Rado, S. (1953) Dynamics of classification of disordered behavior. *American Journal of Psychiatry* 110:406–416.

Rao, D. C., Morton, M. E., Gottesman, I. I. & Lew, R. (1981) Path analysis of qualitative data on pairs of relatives: Application to schizophrenia. *Human Heredity* 31:325–333.

Reich, T., Rice, J. & Cloninger, C. R. (1981) The detection of a major locus in the presence of multifactorial variation. In *Genetic Research Strategies for Psychobiology and Psychiatry*, ed. E. S. Gershon, S. Matthysse, X. O. Breakefield & R. D. Ciaranello, pp. 353–367. Pacific Grove, CA: Boxwood Press.

Reich, W. (1976) The schizophrenia spectrum: A genetic concept. *Journal of Nervous and Metal Disorders* 162:3–12.

Rieder, R. O. & Nichols, P. L. (1979) Offspring of schizophrenics. III. Hyperactivity and neurological soft signs. *Archives of General Psychiatry* 36:665–674.

Reveley, A. M., Erlenmeyer-Kimling, L., Holzman, P. S., Karlsson, J. L., Kendler, K. S., Kety, S., et al. (1987) Genetics as an approach to etiology. In *Biological Perspectives in Schizophrenia*, ed. H. Helmchen & F. A. Henn. New York: John Wiley.

Risch, N. & Baron, M. (1984) Segregation analysis of schizophrenia and related disorders. *American Journal of Human Genetics* 36:1039–1059.

Rosanoff, A. J. (1911) A study of heredity in insanity in light of the Mendelian theory. *American Journal of Insanity* 68:221–261.

Rosenthal, D. (1960) Confusion of identity and the frequency of schizophrenia in twins. *Archives of General Psychiatry* 3:297–304.

Rosenthal, D., Wender, P. H., Kety, S. S., Schulsinger, F., Welner, J. & Ostergaard, L. (1968) Schizophrenics' offspring reared in adoptive homes. In *The Transmission of Schizophrenia*, ed. D. Rosenthal & S. S. Kety. Oxford: Pergamon Press.

Rudin, E. (1916) *Zur Vererbung und Neuentstehung der Dementia Praecox*. Berlin: Springer.

Saitoh, O., Niwa, S., Hiramatsu, K., Kameyama, T., Rymar, K. & Itoh, K. (1984) Abnormalities in late positive components of event-related potentials may reflect a genetic predisposition to schizophrenia. *Biological Psychiatry* 19:293–303.

Scharfetter, C. & Nusperli, M. (1980) The group of schizophrenias, schizoaffective psychoses and affective disorders. *Schizophrenia Bulletin* 6:586–591.

Schulsinger, H. (1976) A ten-year follow up of children of schizophrenic mothers: Clinical assessment. *Acta Psychiatrica Scandinavica* 53:371–386.

Sherrington, R., Brynjolfsson, J., Petursson, H., Potter, M., Dudleston, K., Barraclough, B., et al. (1988) Localization of a susceptibility locus for schizophrenia on chromosome 5. *Nature* 336:164–167.

Siever, L. J. & Gunderson, J. G. (1983) The search for a schizotypal personality. *Comprehensive Psychiatry* 24:199–212.

Slater, E. (1953) *Psychotic and Neurotic Illnesses in Twins*. London: Her Majesty's Stationery Office.

Spitzer, R. L., Endicott, J. & Gibbon, M. (1979) Crossing the border into borderline personality and borderline schizophrenia: The development of criteria. *Archives of General Psychiatry* 36:17–24.

Spitzer, R. L., Endicott, J. & Robins, E. (1978) Research diagnostic criteria: Rationale and reliability. *Archives of General Psychiatry* 35:773–782.

Sturt, E. & McGuffin, P. (1985) Can linkage and marker association resolve the genetic aetiology of psychiatric disorders? Review and argument. *Psychological Medicine* 15:455–462.

Tienari, P. (1971) Schizophrenia and monozygotic twins. *Psychiatria Fennica*, 97–104.

Tienari, P. (1975) in Finnish male twins. In *Studies of Schizophrenia*, ed. M. H. Lader. Ashford, Kent: Headley Brothers.

Torgersen, S. (1984) Genetic and nosological aspects of schizotypal and borderline personality disorders: A twin study. *Archives of General Psychiatry* 41:546–554.

Tsuang, M. T., Bucher, K. D. & Fleming, J. A. (1982) Testing the monogenic theory of schizophrenia: An application of segregation analysis to blind family study data. *British Journal of Psychiatry* 140:595–599.

Tsuang, M. T., Bucher, K. D. & Fleming, J. A. (1983) A search for schizophrenia spectrum disorders: An application for a multiple threshold model to blind family study data. *British Journal of Psychiatry* 143:572–577.

Tsuang, M. T., Winokur, G. & Crowe, R. R. (1980) Morbidity risks of schizophrenia and affective disorders among first-degree relatives of patients with schizophrenia, mania, depression and surgical conditions. *British Journal of Psychiatry* 137:497–504.

Turner, W. J. (1979) Genetic markers for schizotaxia. *Biological Psychiatry* 14:177–205.

Vogler, G. P., Gottesman, I. I., McGue, M. K. & Rao, D. C. (1990) Mixed model segregation analysis of schizophrenia in the Lindelius Swedish pedigrees. *Behavior Genetics* 20:461–472.

Weintraub, S. (1987) Risk factors in schizophrenia. *Schizophrenia Bulletin* 13:439–450.

Wender, P. H. & Klein, D. F. (1981) *Mind, Mood & Medicine*. New York and Scarborough, Ontario: Meridian.

Wender, P. H., Rosenthal, D., Kety, S. S., Schulsinger, F. & Welner, J. (1974) Cross-fostering: A research strategy for clarifying the role of genetic and experimental factors in the etiology of schizophrenia. *Archives of General Psychiatry* 30:121–128.

Zilboorg, G. (1941) Ambulatory schizophrenias. *Psychiatry* 4:149–155.

Chapter 8
Epidemiology and Genetics of Depression and Mania

GEORGE WINOKUR

The new era in genetic and epidemiologic studies in the affective disorders (depressions and mania) dates from 1966, at which time manic-depressive disease (bipolar) was separated from depressive disease (unipolar) by three groups of investigators working independently in Sweden, Switzerland, and the United States. In fact, Leonhard had suggested a monopolar-bipolar dichotomy some years earlier (1957). As time evolved, it became clear that bipolar illness was characterized by a different family history, a different age of onset, and an increased frequency of affective episodes, as well as by the presence of mania in the clinical picture. The major differentiating factors are presented in Table 8.1 (Winokur, 1978). This separation of bipolar from unipolar illness constitutes the point of departure for this chapter.

Very distinguished family and twin studies had been done prior to 1966. However, the separation of bipolar from unipolar illness made it necessary either to reevaluate these studies or to initiate a new series of investigations; it was clear that the two illnesses might be independent. Further, there were problems in diagnoses. Early work by Cassidy et al. (1957) had indicated the feasibility of using a set of criteria for the diagnosis of depression. Both for studies of families of affectively ill probands and for the determination of the frequency of manias and depressions in the general population, it became necessary to have systematic criteria that could be agreed upon by all who were involved in genetic and epidemiological investigations. The last 10 to 15 years have been marked by important new advances in the diagnoses of the affective syndromes, as well as by new genetic and epidemiological findings in bipolar and unipolar illnesses.

TABLE 8.1.
Familial and Clinical Differences between Manic-Depressive
Disease (Bipolar) and Depressive Disease (Unipolar)

Familial	Bipolar	Unipolar	P
Affective disorder in a parent	52%	26%	0.001
Affective illness in a parent or extended family	63%	36%	0.0005
Mania in first-degree relative	3.7%–10.8%	0.29%–0.35%	
Clinical			
Presence of mania	Yes	No	
Bi- or triphasic immediate course	Yes	No	
4 or more episodes after index admission	18%	6.5%	
Proportion female	59%	64%–68%	
Hypersomnia	More	Less	
Diurnal variation	More	Less	
Pacing behavior	Less	More	
Anger	Less	More	
Somatic complaints	Less	More	

Systematic Criteria and the Epidemiology of Depressions and Manias

The Feighner Criteria (Feighner et al., 1972) were the first of several extensive sets of systematic diagnostic criteria formulated for use in research. Besides specific symptoms necessary for a diagnosis of depression and mania, these criteria present time constraints: for depression, one month; for mania, two weeks. If the individual does not meet the specific symptom criteria and time constraints, he may be called undiagnosed. Research Diagnostic Criteria (RDC), developed by Spitzer et al. (1978), succeeded the Feighner Criteria. The symptom requirements bear similarities to the Feighner Criteria, but for a diagnosis of definite depression, only two weeks of illness are needed; only one week is necessary for a diagnosis of probable depression or for a diagnosis of mania. Though the two sets of criteria are similar, these are substantial differences. Both sets have been used for epidemiologic studies.

Other criteria used are those of the *Diagnostic and Statistical Manual*, 3rd edition (DSM -III) (1980), published by the Task Force on Nomenclature and Statistics of the American Psychiatric Association. The diagnostic symptoms described in the DSM-III relate to the two previous sets of diagnostic criteria: the duration for mania in DSM-III is one week, and for depression, at least two weeks. There is a special category for those depressives who have "melancholia." This involves the loss of pleasure in all or almost all activities and a lack of reactivity to usually pleasurable stimuli. Also, there is a category called "dysthymic disorder," which is used for individuals with relatively mild symptomatology but relatively persistent depressive syndromes. In 1987, these criteria were revised. DSM-III-R (1987) contains the same two-week criteria for depression but the time constraint for mania (one week) is not present.

The use of diagnostic criteria became fashionable because it was clear that they improved reliability and diagnosis. However, after the passage of time, reliability between two diagnosticians is likely to go down. A clinician making an assessment at time of admission may not necessarily agree with a clinician making an assessment six months later. This may result from a change in the clinical state, forgetfulness, or differences in what each clinician perceives to be a positive response. Further, even using the criteria, serious diagnosticians may disagree because of differences in judgment about the presence of specific symptoms, as well as differences in interpretation of the criteria (Winokur et al., 1988).

The use of systematic criteria does not obviate the need for differential diagnosis. For example, it is conceivable that an individual would be diagnosed as having a depression or mania if he/she did not meet the criteria. On the other hand, having met the criteria does not mean that an appropriate diagnosis has been made. To make an appropriate diagnosis, it is absolutely necessary to subject an individual's clinical picture to a differential diagnosis. Even here a problem remains: all of the criteria have been formulated on the basis of studies of psychiatric patients. A significant doubt remains whether people in the community who meet the criteria in fact have the illness. This will become apparent as we discuss some of the population studies.

In a large epidemiologic and genetic study—the Iowa 500

TABLE 8.2.
Affective Disorder in 1,578 Blindly and Personally Examined
First-Degree Relatives of Schizophrenia, Mania, Depression,
and Control Diagnostic Groups

Diagnostic group	N	#AD (%)	BZ	MR (%) ± S.E.
Schizophrenia (S)	354	19 (5.4)	273	7.0 ± 1.54
A.S.	375	29 (7.7)	287	10.1 ± 1.78
Mania (M)	216	21 (9.7)	160	13.1 ± 2.65
A.S.	230	30 (13.0)	169	17.8 ± 2.94
Depression (D)	467	44 (9.4)	341	12.9 ± 1.82
A.S.	500	66 (13.2)	362	18.2 ± 2.03
Control (C)	541	26 (4.8)	344	7.6 ± 1.43
A.S.	543	27 (5.0)	345	7.8 ± 1.45

NOTE: Risk period 15–59.
Pairwise comparisons for personal examination material: S:M*, S:D*, S:C, M:D, M:C+,
D:C*.
$+p < .10$; $*p < .05$
Italic numbers refer to all sources data (A.S.) and include information from records on
family members identified as ill by other family members, as well as personal examination
(Total N = 1,648).
Pairwise comparisons for all sources material: S:M*, S:D*, S:C, M:D, M:C*, D:C*.
$*p < .05$

Project—all available and living first-degree relatives were personally examined and diagnosed (Tsuang et al., 1980). Table 8.2
shows the morbidity risks for affective disorder (mania and/or
depression) in relatives of 200 schizophrenic probands, 100 manic
probands, 225 depressive probands, and 160 surgical controls.
The relatives of the controls may be considered as representative
of the general population. Any control with a suspected psychiatric illness was eliminated from the study. This decision to
eliminate psychiatrically ill controls was made on the basis of
notes recorded on the surgical charts; those eliminated were few
in number.

According to the results of personal interviews, the morbidity
risk for affective disorder in family members of controls is 7.6%.
Morbidity risk (or disease expectancy) is an estimate of the proportion of the population that will develop the disease in
question, given a survival of the risk period (manifestation
period). Because mortality rates increase for those with schizophrenia and affective disorder, some of these relatives could not
be interviewed (Tsuang & Woolson, 1977). However, records

were obtained and this material is recorded as A.S., or all sources material. The number of first-degree relatives studied goes up from 1,548 to 1,648, but the morbid risk for control relatives increased from only 7.6% to 7.8%.

These data should be compared to the material from the New Haven Study, which used the Research Diagnostic Criteria (Weissman & Myers, 1978). The subjects in this study were chosen at random from 1,095 households, and 511 adult subjects were interviewed and diagnosed. Table 8.3 presents the lifetime rates for major depression, minor depression, and bipolar illness. Had a morbid risk formula been used, the rates in the New Haven Study would have been higher. It is possible that some of the patients who were diagnosed as having minor depression in the New Haven Study might not have been diagnosed as having depression in the Iowa 500 because they might have had too few symptoms. However, the 20% of patients diagnosed with major depression would have had enough symptoms for a diagnosis of depression in the Iowa 500. To this should be added bipolar patients (1.2%). This total of 21.2% had either major depression or bipolar illness, probable or definite. This should be compared with the results of personal interviews in the Iowa 500 Project, where 7.6% of relatives showed an affective disorder.

In the three categories listed above (major and minor depression, bipolar illness), the New Haven Study departs from the Iowa 500 Project data. A possible explanation for this is the research criteria used in each study. The majority of major depressions in the New Haven Study were primary, 86%. However, 14% of them were secondary. This means that, prior to the

TABLE 8.3.
Lifetime Rates (Rates/100) of Affective Disorders
(Probable and Definite) in the New Haven Population

	Lifetime rates (%)
Major depression	20.0
Minor depression	9.2
Major and/or minor	26.7
All bipolar	1.2
Total bipolar & unipolar	27.9

time the patient showed a major depressive episode, he had suffered from another disorder such as schizophrenia or alcoholism.
All of the Iowa 500 affective disorder diagnoses were made on the
basis of primary depression or mania. The time constraint for the
Iowa 500 is four weeks; for the New Haven Study, one to two
weeks. A time constraint of four weeks would reduce the number
of patients who were given the diagnosis of primary major depression in the New Haven Study by a considerable amount.
These are important differences to consider in evaluating either
result.

Intuitively, one questions whether finding that 20% to 30% of
people in the community had an affective disorder is meaningful.
In genetic studies the comparison between controls and relatives
of probands must be made on the basis of primary diagnoses. If a
genetic factor were present, there would be a clear increase of primary affectively ill family members of bipolars and unipolars
when they were compared to controls. The inclusion of a variety
of other disorders, which happened from time to time to manifest
themselves as affectively ill (secondary depressives), would not
provide an appropriate comparison. Further, short-lived affective
symptoms may not be a manifestation of significant illness. On
the other hand, for purposes of evaluating a clinical problem in
the community, for example, the presence of depressive or manic
episodes, regardless of their clinical meaning, the New Haven
Study might well be more useful than a study such as the Iowa
500. Both studies, however, suffer from a major problem: respon-
dents may forget about clinical material, thereby affecting
evaluations of lifetime prevalence rates or morbid risks.

A paper by Jenkins et al. (1979) revealed that within a 9-month
period 34% to 46% fewer life events were remembered than occurred. These authors reviewed the literature and remarked that
even hospitalizations were underreported by as much as 50% for
a period of nearly 12 months before the recall interview. An example of this may be seen in the New Haven Study as it presents
the grief material. Grief or bereavement in this study met the criteria for major depression, but was only diagnosed if the depression occurred after the death of someone close. The lifetime rates
were 2.7% for men and 16.2% for women. Clearly, there is an excess of women who show a major depressive episode in response
to bereavement.

In a similar study, Clayton's group reported on the depression of widowhood (Bornstein et al., 1973). At 1 month after a death, 36% of females and 33% of males were diagnosed as having a grief reaction (a depression). At 13 months the incidence was 17% in females and 19% in males. Two things stand out in this comparison: the Clayton material, which deals with the time of bereavement, shows a far higher incidence of depression; further, there is no evidence of any differential between sexes. It is possible that the New Haven Study, which depends on memory, is not correct because of the simple process of forgetting. Moreover, if forgetfulness is present, it may be differentially expressed with men forgetting more than women.

These are problems affecting all studies that attempt to give a lifetime prevalence or morbid risk. Thus, both in the Iowa 500 and New Haven Study, it is quite possible that the rates would be higher if it were possible to deal with the problem prospectively. Probably the only way to deal with this problem in a genetic study is to have an adequate control group. Even then one must assume that the amount of forgetting or nonreporting is equal both in the proband and their relatives, as well as in the controls and their relatives. Though this is a reasonable assumption, it should be tested.

The best-known epidemiologic study in recent years was published in 1984 (Robins et al., 1984). This study used a new diagnostic interview schedule, which was administered by lay interviewers. DSM-III was used as the criteria for psychiatric diagnoses. The study was done in three separate epidemiologic catchment area sites (New Haven, CT; Baltimore, MD; St. Louis, MO). In all, 9,543 people were evaluated. Table 8.4 gives the range of diagnoses for the affective disorders. These lifetime prevalences differ between centers. Lifetime prevalence identifies the proportion of persons in a sample of the population who have ever experienced the disorder in question up to the date of assessment. It should be noted that in Baltimore the lifetime prevalence of mania is 0.6% but in St. Louis and New Haven it is 1.1%, almost twice as high. Major depressive disorder is lowest in Baltimore (3.7%), higher in St. Louis (5.5%), and highest in New Haven (6.7%). New Haven and Baltimore are significantly different, as are St. Louis and Baltimore. Dysthymia is highest in St. Louis and lowest in Baltimore. New Haven versus Baltimore is statistically significant, as is St. Louis versus Baltimore.

If one looks at all affective disorders, the lifetime prevalence is

TABLE 8.4.
Lifetime Prevalence Rates of Affective Disorders
in Three Epidemiologic Catchment Area Sites

Manic episode	0.6%–1.1%
Major depressive episode	3.7%–6.7%
Dysthymia	2.1%–3.8%
Affective disorders	6.1%–9.5%

9.5% in New Haven, 6.1% in Baltimore, and 8.0% in St. Louis.
The difference between Baltimore and New Haven is more than
50% based on the lower prevalence rate. What may be said is that
ever increasing specificity of diagnosis and ever increasing precision of interviewing have not totally solved the problems of
epidemiology. Of course, one thing is quite possible; there may,
in fact, be 50% more affective illness in New Haven than there is
in Baltimore. That would have to be proven, but it remains a legitimate possibility.

Genetic Studies of Manic-Depressive Illness (Bipolar)

The presence of mania in the proband is the nuclear difference
between bipolar and unipolar illness. Equally important is the
high incidence of mania in the relatives of bipolar patients. In an
extensive clinical and family study of manic-depressive illness
(bipolar), Winokur et al. (1969) reported a very high set of morbidity risks; 34% ± 4.6% for parents and 35% ± 5.0% for siblings.
These data came from "all sources" (they reflected information
from personal examinations, records, family history, reports
from physicians, etc.). In fact, in this study 167 first-degree relatives were personally interviewed. The morbidity risks for these
relatives were 23% for males and 56% for females. There was clear
evidence that within a bipolar family, the illness did not breed absolutely true. Thus 29 of the first-degree relatives had only
depression, whereas 17 had suffered a mania. Depression only is
the modal illness within the family of a bipolar. Assuming that in
these families both the depressives and the bipolars have the

same illness, the data in this study suggest a dominant gene transmission.

Both a control group and blindness in the assessment of familial illness are necessary for an evaluation of the family. The Iowa 500 Project started with defining a group of schizophrenics, manics, and depressives according to the Feighner criteria. These people had been admitted in the 1930s and early 1940s. A group of controls (surgical patients) was also selected. All probands and available first-degree family members were personally evaluated 30–40 years later. In addition to hospital records and physicians' reports, some family members provided information about other members of the family (family history method). In general, to evaluate family rates among the affectively ill we may use two control groups, the family members of the schizophrenics and the family members of the controls. Both control groups showed a comparable amount of affective disorder, which should be lower than what would be found in the family members of the affectively ill probands. In the Iowa 500 study, the interviews were done blind to the diagnosis of the proband (Tsuang et al., 1980). Most of the comparisons between the control groups and the bipolar and unipolar relatives are significantly different (see Table 8.2). Of interest is the fact that the expected increase in familial illness in the manics, as opposed to the unipolar depressives (found in other studies), was not found in the personal examination portion of the material.

An evaluation was then made of the all sources material (A.S.). It is noteworthy that the amount of illness goes up considerably in the family members of manics and depressives, but it does not change as much in the family members of the controls or schizophrenics (see Table 8.2). When the all sources material is used, comparisons between the psychiatrically well control groups and depressives are significantly different from each other. Again, there is no increase in affective illness in the family of manics over affective illness in the families of depressives, a surprising finding in light of previous studies.

There are interesting methodological considerations in this study. Prior to engaging in the field work of the Iowa 500, a blind family history assessment was done on the records of the 525 psychiatrically ill probands of the Iowa 500 (Table 8.5). The charts contained a considerable amount of material on the family history. These family members were evaluated blind to the diagnosis

TABLE 8.5.
Morbidity Risks for Affective Disorder in Parents and Sibs
of Manic, Depressive, and Schizophrenic Probands

	# with affective disorder	# at risk	Morbidity risk (%)
Schizophrenic probands (N = 200)			
Parents and sibs	32	584	5.5
Manic probands (N = 100)			
Parents and sibs	37	328	11.3
Depressive probands (N = 225)			
Parents and sibs	123	859	14.3

NOTE: Pairwise comparisons: S:M*, S:D*, M:D
*$p < .005$

of the proband (Winokur et al., 1972). There was no control group, but if one considers the schizophrenic probands as a control group, clear differences emerge. Strikingly significant increases in affective psychopathology in the families occur for the manic probands, as well as for the depressive probands. Again, the difference between the bipolars and unipolars is not significant. If one compares this table (Table 8.5) with the material from the field work, family history, and personal examination evaluation (Table 8.2, the all sources data), it seems clear that the morbidity risks are lower in the simple family history workup. Significantly, the family history method provides relative differences in the probands versus schizophrenic controls similar to the family study method (personal interview). There is no doubt that for absolute values, personal examinations and other sources of information must be used. However, for purposes of evaluating relatives and comparing groups within the same study, the family history seems adequate.

Twin research conducted in Denmark by Bertelsen et al. (1977) indicates that genetic factors are present in bipolar illness. Bertelsen and his co-investigators selected their sample from the Danish Psychiatric Twin Register and the Danish Central Psychiatric Register. Therefore, their sample consists of systematically ascertained psychiatric index cases from an unselected twin population of same sex twins born in Denmark in the years 1870 to

TABLE 8.6.
Pairwise Concordance* Rates in Unipolar
and Bipolar Twin Pairs

	MZ	DZ
Unipolar	12/28 (43%)	3/16 (19%)
Bipolar	20/27 (74%)	6/36 (17%)

SOURCE: From Bertelsen, Harvald & Hauge (1977)
*Strict concordance, when co-twin had diagnosis of manic depressive disorder, certain or probable without regard to polarity or severity.

1920. Table 8.6 represents the concordance rates that were published. It is noteworthy that in the bipolar twins, 74% of monozygotic twins were concordant, as opposed to 17% in the dizygotic group. This is certainly in the direction one would expect if a significant genetic factor were operating. Fully a third of the concordant bipolar pairs were not concordant for bipolar illness; rather, they were composed of one bipolar or manic twin and the other a unipolar twin. Thus, even among twins, bipolar illness does not breed absolutely true. Among many of concordant twin pairs, one of the twins has only depression.

An unequivocal genetic factor in psychiatric illness may be demonstrated by use of an adoption study. Mendlewicz and Rainer (1977) have presented important data showing a clear genetic factor in bipolar illness through the use of the adoption study methodology. In 29 adoptees who had bipolar illness, 31% of biological parents had an affective illness, as opposed to only 12% of adoptive parents ($p < .025$). There were a variety of control groups. Of 31 bipolar nonadoptees, 26% of parents had an affective illness; of 22 normal adoptees, 10% of adoptive parents, and 2% of biological parents, had an affective illness. The biological parents of 20 poliomyelitis patients showed only 10% affective disorders. Whether the person is adopted or not seems irrelevant. The biological parent of a bipolar patient, adopted or not, runs a high probability of having an affective disorder. Such findings as these remove all doubt about the presence of a genetic background in bipolar illness.

Finally, an entire area of research in bipolar illness is concerned with linkage between a presumed genetic locus for this illness and known genetic markers. If two genes have their loci on the same chromosome and are close to each other, they do not assort independently; therefore, the loci are said to be linked. X-linkage was proposed in bipolar illness in the late 1960s. The reasons for considering the possibility of X-linkage were the findings that, within a family of a bipolar patient, more females than males were affected with affective illness and there was no male:male (father to son) transmission. Additionally, in two informative families, it was possible to trace a particular marker, for example, the color blindness trait. Within these families affective illness went along with this X-linked gene (Winokur et al., 1969). This did not mean that color blindness was associated with bipolar illness; rather it indicated that in the two families a known marker was useful as a tracer. The striking findings favoring X-linkage were the lack of male:male transmission and the assorting of a genetic marker (color blindness) with the affective illness. As males do not give an X-chromosome to their sons, the lack of male:male transmission would be highly in favor of X-linkage. As the color blindness loci are on the X-chromosome, the association within a family of a specific marker and bipolar illness would indicate X-linkage. Mendlewicz and Rainer (1974), in a study of 134 manic-depressive probands, also found family data that was compatible with an X-linked dominant mode of transmission.

Mendlewicz and Fleiss (1974) collected 17 informative kindreds and found linkage between both deutan and protean color blindness. Less close linkage was found between bipolar illness and the Xga blood system. Baron (1977) described a large family that demonstrated statistically measurable linkage between color blindness and manic-depressive illness. In this family, the proband was schizoaffective, suggesting that schizoaffective disorder and manic-depressive illness run in the same families. This was the first time that such a finding had occurred in a linkage study, although a few years before McCabe et al. (1971), in a blind family study, had shown a marked relationship between affective illness and "good prognosis schizophrenia" (a synonym for schizoaffective illness).

Gershon and Bunney (1976) evaluated the question of X-linkage in bipolar illness. They concluded that some of the reported families were subject to alternate analyses and might not be informative. Further, they pointed out that close linkage of bipolar illness to both the Xga blood system locus and the color blindness loci was not compatible with what was known about the map distance between the Xga locus and the color blindness loci. The Xga locus and the color blindness loci were too far apart for a presumed locus for manic-depressive illness to be linked to both of them. Gershon et al. (1979) have reported on six families that were assorting for both color blindness and bipolar illness. No evidence for linkage between the color blindness loci and bipolar illness was found.

The question of the male:male transmission has now been investigated in a variety of studies (Goetzl et al., 1974; Helzer & Winokur, 1974) and such transmissions have been demonstrated conclusively in bipolar illness. Mendlewicz et al. (1980) studied a large kindred in which linkage was found between G6PD and bipolar illness. G6PD is another X-marker, one not too far from the color blindness loci. A legitimate possibility exists that more than one kind of transmission occurs in bipolar illness: X-linkage and autosomal. Therefore, some families would show X-linkage and others would not. This may also explain the finding of male:male transmission. It would be the non-X-linked families where such a finding would occur. Such heterogeneity in transmission is not unknown in biology (e.g., in retinitis pigmentosa).

Recently, there has been a spate of new linkage findings in bipolar illness. Baron and his co-workers (1987) reported data from a pedigree study showing close linkage of bipolar affective illness to X-chromosome markers, such as color blindness and G6PD. The findings were very striking, with lod scores ranging from 7.52 to 9.26. Thus, the odds favored linkage very heavily, between 10^{-7} and 10^{-9}.

Other major contributions have come out of the molecular genetics laboratory. Mendlewicz and his co-workers (1987) have presented evidence in favor of linkage between manic-depressive illness and coagulation factor 9 at the Xq27 region of the X-chromosome in DNA samples from peripheral leucocytes of manic-depressive probands and relatives in 10 informative families. Interestingly, the F9 locus may be distant from the color blindness locus, suggesting two X-chromosome loci for manic-depressive disease.

Egeland et al. (1987) studied a very large pedigree among the Amish. She and her co-workers localized a dominant gene for manic-depressive illness (bipolar), conferring a strong predisposition to manic-depressive disease to the end of the short arm of chromosome 11. The markers that are specifically cited as linked are the insulin gene and the Harvey-ras gene. However, attempts to look for linkage in the same area have not been successful in other laboratories. Hodgkinson and his co-workers (1987) studied three families and were unable to find linkage between the Harvey-ras gene or the insulin gene and manic-depressive illness. These pedigrees came from Iceland, and it did appear that a single autosomal dominant disease allele was segregating. They concluded that this shows evidence of genetic heterogeneity in manic depression, as there was no evidence in favor of linkage with the markers that were reported by Egeland et al. (1987). Detera-Wadleigh and her co-workers (1987) also looked at the question of linkage between the Harvey-ras gene and the insulin gene in three families but were unable to find evidence for linkage between these markers and a locus for affective disorder.

A reasonable conclusion from these studies is that there are at least three to four different genetic loci (and possibly more) that are involved in bipolar illness. One or more of these is on the X-chromosome. Another is on chromosome 11, and the other possible areas of linkage are unknown. However, many areas may be involved in linkage. Those pedigrees that do not fit either X-linkage or chromosome 11 linkage may have a variety of other sites that are transmitting the illness.

Though bipolar illness appears to be a relatively homogeneous entity, some data suggest the possibility of various etiological types. Krauthammer and Klerman (1978) have reviewed the literature and found good evidence for the presence of a "secondary mania," which would follow, or be related to, certain kinds of medical circumstances, such as corticosteroid administration and neoplasm. It is quite possible that some patients who enter a study for bipolar illness are in fact secondary manics in this sense. Further, a paper by Kadrmas and Winokur (1979) indicates that bipolar patients who have an abnormal EEG are less likely to have a positive family history for affective disorder than bipolar patients with a normal EEG. This has been noted before and would indicate a possibility of heterogeneity within the bipolar syndrome. It also raises the possibility that those patients with an abnormal EEG suffer from an illness that is polygenic in nature; the abnormal EEG may be related to an increasing susceptibility that would make fewer genes necessary.

A study by Kadrmas et al. (1979) shows that postpartum mania is associated with more delusional symptoms and a better short-term prognosis for fewer episodes than is found in nonpostpartum manic women. These findings, which suggest the possibility of heterogeneity within the bipolar syndrome, should be appraised in future investigations.

Family Data in Unipolar Depression

We recall that in Table 8.2 and Table 8.5, familial affective pathology is more frequent in depressive probands than in the control groups of both normals and schizophrenic probands.

Bertelsen et al. (1977) show an increased concordance rate in monozygotic unipolar twin pairs as opposed to dizygotic unipolar pairs. The difference is not as great as those in the bipolar pairs. The fact that 74% of the pairwise comparisons in the monozygotic pairs in bipolars are concordant, as opposed to 43% in the unipolar pairs, is of considerable interest. This is a significant difference ($p < .05$) and has been interpreted to indicate that unipolar depression is a less genetic illness than bipolar illness. A perfectly legitimate interpretation is that the unipolar depression is composed of more than one illness. If this were true and some illnesses had a significant genetic factor and some did not, it would account for the lower rate of concordance in the unipolar pairs. It is of interest that the dizygotic pairwise concordances do not differ (the unipolar at 19% and the bipolar at 17%). One would expect the bipolar to show a higher concordance rate than the unipolar in the dizygotic pairs based on the two monozygotic concordance rates. The numbers of pairs and concordances are relatively small; perhaps it is too much to expect to find such a difference under these circumstances. However, there is not even a trend and if, in fact, unipolars are less genetic or composed of a heterogeneous group of patients, one would expect to find a differential between bipolar and unipolar dizygotic pairs. Hence, the similarity in the dizygotic pairs remains unexplained.

There are two adoption studies that are relevant to unipolar depression. Cadoret (1978) found that two of the three adoptees of unipolar depressed biological parents themselves had unipolar depression (67%). In two control groups (N = 118), eight (7%) had a depression. This is a significant difference. Wender et al. (1986) showed more familial affective disorder (4.5%) in unipolar depressives plus neurotic depressives when compared to familial affective disorder in controls (2.3%). Familial alcoholism was significantly higher in a mixed set of affective probands (bipolar + unipolar + patients with an affective reaction) when compared to familial alcoholism in controls, 5.4% vs 2%).

Unipolar depression is a highly prevalent syndrome; it seems reasonable to consider the possibility that it is also a heterogeneous syndrome. There have been attempts to separate out subgroups of unipolar depression based on family background. Winokur et al. (1971) separated three groups. The first of these was called "depression spectrum disease." It was an ordinary depression seen in an individual who had a family history of alcoholism. Such an individual might or might not have a family history of depression. Specifically, a family background of alcoholism was the marker for the illness. The second group was familial pure depressive disease (Winokur et al., 1978). This is an ordinary depression in an individual who has a first-degree family history of unipolar depression in the family but no family history of alcoholism, antisocial personality, or mania. A final group is called sporadic depressive disease. This group has no first-degree family history of alcoholism, antisocial personality, mania, or depression. In this separation of three groups, there is no concern with whether there is a genetic factor or what kinds of transmissions might be involved. However, there is present the assumption that different kinds of familial transmission (possibly genetic) are identifying different kinds of patients. This assumption is used to determine autonomous subtypes of unipolar depression.

Some interesting findings arose from a comparison of familial pure depression versus depression spectrum disease (see Table 8.7 for a summary of differences). It is clear that the former is more likely to have had a previous admission and also to have more hospitalizations and episodes in a follow-up (Winokur, 1983). It is also clear that both familial pure depressive disease and depression spectrum disease have an age of onset in the early 30s. Sporadic depressive disease has its onset significantly later than the previous two types, in the late 30s to early 40s. There is a dramatic difference in an endocrinological measure, the dexamethasone suppression test (Schlesser et al., 1979). Very few (4%) of the depression spectrum disease patients have abnormal nonsuppressor status, but 82% of the familial pure depressives

TABLE 8.7.
Familial Subtypes of Primary Unipolar Depression

	Familial pure depressive disease	Depression spectrum disease	Sporadic depressive disease
Family history of mania	No	No	No
Family history of depression	Yes	Yes	No
Family history of alcoholism	No	Yes	No
Age of onset	30s	30s	40s
Stormy personal life	No	Yes	No
Nonsuppressor on dexamethasone suppression test	82%	4%	43%
Multiple episodes	High	Low	Middle

are abnormal nonsuppressors. Not all studies have shown this difference but many have. The sporadics show 43% abnormal nonsuppression. It has been clear from previous family studies that such patients as the sporadics might in fact contain a large number of false negatives for family members. In family studies or family history studies one sees many false negatives but few false positives. Therefore, the depression spectrum and the familial pure depressive patients are more reliably ascertained; the sporadics probably contain considerable errors because of these possible false negatives. It is of interest that both course of illness and the dexamethasone suppression test differentiate familial pure depressives from depression spectrum patients. This would indicate that a familial method for separating unipolar depressives may well be worthwhile and may identify autonomous subgroups.

The concept of an illness such as depression spectrum disease in psychiatry poses major problems for definition. Essentially what is being stated is that an illness expresses itself differently, sometimes by alcoholism and sometimes by depression; there may be a genetic factor that accounts for the propensity to develop one or the other of these clinical syndromes. If this is the case, adoption studies might shed some light on the concept. Goodwin et al. (1977) investigated adoptees whose parents had alcoholism. Women born of alcoholic parents and raised by their biological alcoholic parents showed a significant increase in depression over controls, but women of alcoholic parents who were brought up by adopted parents did not show any increase in depression. This might be interpreted as indicating that only an environmental factor is involved or that there is some kind of interaction between an environmental and genetic factor. Von Knorring et al. (1983) investigated a set of adoptees. She and her co-workers found a significant excess of biologic mothers with substance abuse in depressive adoptees. Because this study made multiple comparisons, it was conceivable that the finding was simply due to the large number of comparisons, one or more of which might have turned out significant by chance alone. An adoption study that bears on the same question is that of Wender et al. (1986). He and his colleagues studied depressed adoptees (bipolar, unipolar, and affective reaction patients) and found that there was an increase in alcoholism in the biologic relatives of those depressive adoptees versus control adoptees. In fact, then, all studies show a relationship between alcoholism and depression in relatives who have been separated by adoption. The question of the role of environment is as yet unknown. There is a suggestion, however, that a genetic factor does exist that is relevant to a type of unipolar depression called depression spectrum disease.

Linkage studies in depression spectrum disease and familial pure depressive disease have been relatively few, but Hill and co-workers (1988), as well as Tanna et al. (1989), have shown evidence for linkage using the marker orosomucoid on chromosome 9. Essentially what has been shown is that in families where unipolar depression and alcoholism coexist, these illnesses are linked to orosomucoid locus, which occurs on chromosome 9. Several families with familial pure depression were also studied, and it appeared that the MNS locus might be linked to a locus for familial pure depressive disease. All of these studies are preliminary, but they are suggestive. If replicated, they would show a clear and unequivocal genetic background for depression spectrum disease and familial pure depressive disease.

Conclusion

A major new methodology in psychiatry has been the development of systematic criteria for illness. This is certainly a step forward but the criteria may identify too many people in the general population as being ill. This possibility constitutes a problem for quantitative genetic studies.

New genetic data in bipolar illness indicate strong findings in favor of a genetic background. These findings have emerged from blind family studies comparing bipolar probands with controls, from twin studies, and from an adoption study. X-linkage has been investigated and there is evidence for its existence in bipolar illness. It is reasonable to suppose that there is more than one kind of transmission in bipolar illness, X-linked and autosomal, and there may be several loci for it.

Unipolar illness, likewise, appears familial: family studies and twin studies indicate a genetic factor. Some adoption studies suggest a genetic factor in unipolar depression but more data are needed. A new methodology employs the use of family background to separate out specific types of unipolar illness. This methodology shows some promise. Separation by family background has indicated three groups, two of which, depression spectrum disease and familial pure depressive disease, seem quite different from each other both in the course of an illness and in a physiological (endocrinological) response. Adoption studies and linkage studies suggest that depression spectrum disease may be a true diagnostic entity.

References

American Psychiatric Association (1980) *Diagnostic and Statistical Manual of Mental Disorders*, 3rd ed. Washington, DC: American Psychiatric Association.

American Psychiatric Association (1987) *Diagnostic and Statistical Manual of Mental Disorders*, 3rd ed. rev. Washington, DC: American Psychiatric Association.

Baron, M. (1977) Linkage between X-chromosome marker (Deutan color blindness) and bipolar affective illness. *Archives of General Psychiatry* 34:721–725.

Baron, M., Risch, N., Hamburger, R., Mandel, B., Kushner, S., Newman, M., et al. (1987) Genetic linkage between X-chromosome markers and bipolar affective illness. *Nature* 326:289–292.

Bertelsen, A., Harvald, B. & Hauge, M. (1977) A Danish twin study of manic depressive disorders. *British Journal of Psychiatry* 130:330–351.

Bornstein, P. Clayton, P., Halikas, J., Maurice, W. & Robins, E. (1973) The depression of widowhood after 13 months. *British Journal of Psychiatry* 122:S61.

Cadoret, R. (1978) Evidence for genetic inheritance of primary affective disorder in adoptees. *American Journal of Psychiatry* 135:463–465.

Cassidy, W., Flanagan, N. Spellman, M. & Cohen, M. (1957) Clinical observations in manic depressive disease: A quantitative study of 100 manic depressive patients and 50 medically sick controls. *Journal of the American Medical Association* 164:1535.

Detera-Wadleigh, S., Berrettini, W., Goldin, L., Boorman, D., Anderson, S. & Gershon, E. (1987) Close linkage of c-Harvey-ras-1 and the insulin gene to affective disorder is ruled out in three North American pedigrees. *Nature* 325:806–808.

Egeland, J., Gerhard, D., Pauls, D., Sussex, J., Kidd, K. Allen, C., et al. (1987) Bipolar affective disorders linked to DNA markers on chromosome 11. *Nature* 325:783–786.

Feighner, J., Robins, E., Guze, S., Woodruff, R., Winokur, G. & Munoz, R. (1972) Diagnostic criteria for use in psychiatric research. *Archives of General Psychiatry* 26:57–63.

Gershon, E. & Bunney, W. (1976) The question of X-linkage in bipolar manic depressive illness. *Journal of Psychiatric Research* 13:99–117.

Gershon, E., Targum, S., Matthysse, S. & Bunney, W. (1979) Color blindness linkage to bipolar illness not supported by new data. *Archives of General Psychiatry* 36:1423–1430.

Goetzl, U., Green, R., Wybrow, P. & Jachson, R. (1974) X-linkage revisited. *Archives of General Psychiatry* 31:665–672.

Goodwin, D., Schulsinger, F., Knop, J., Mednick, S. & Guze S. (1977) Alcoholism and depression in adopted children of alcoholics. *Archives of General Psychiatry* 34:751–755.

Helzer, J. & Winokur, G. (1974) A family interview study of male manic depressives. *Archives of General Psychiatry* 31:73–77.

Hill, E., Wilson, A., Elston, R. & Winokur, G. (1988) Evidence for possible linkage between genetic markers and affective disorders. *Biological Psychiatry* 24:903–917.

Hodgkinson, S., Sherrington, R., Gurling, H., Marchbanks, R., Reeders, S., Mallet, J., et al. (1987) Molecular genetic evidence for heterogeneity in manic depression. *Nature* 325:805–806.

Jenkins, C., Hurst, M. & Rose, R. (1979) Life changes: Do people really remember? *Archives of General Psychiatry* 36:379–384.

Kadrmas, A. & Winokur, G. (1979) Manic depressive illness and EEG abnormalities. *Journal of Clinical Psychiatry* 40:306–307.

Kadrmas, A., Winokur, G. & Crowe, R. (1979) Postpartum mania. *British Journal of Psychiatry* 135:551–554.

Krauthammer, C. & Klerman, G. (1978) Secondary mania. *Archives of General Psychiatry* 35:1333–1339.

Leonhard, K. (1957) *Aufteilung der endogenen Psychosen*. Berlin:Akademie Verlag.

McCabe, M., Fowler, R., Cadoret, R. & Winokur, G. (1971) Familial differences in schizophrenia with good and poor prognosis. *Psychological Medicine* 1:326–332.

Mendlewicz, J. & Fleiss, J. (1974) Linkage studies with X-chromosome markers in bipolar (manic depressive) and unipolar (depressive) illnesses. *Biological Psychiatry* 9:261–294.

Mendlewicz, J., Linkowski, P. & Willmotte, J. (1980) Linkage between glucose-6 phosphate dehydrogenase deficiency and manic depressive psychosis. *British Journal of Psychiatry* 137:337–342.

Mendlewicz, J. & Rainer, J. (1974) Morbidity risk and genetic transmission in manic depressive illness. *American Journal of Human Genetics* 26:692–701.

Mendlewicz, J. & Rainer, J. (1977) Adoption study supporting genetic transmission in manic depressive illness. *Nature* 268:327–329.

Mendlewicz, J., Simon, P., Sevy S., Charon, F., Brocas, H., Legros, S. & Vassant, G. (1987) Polymorphic DNA markers on X-chromosome and manic depression. *Lancet* May 30:1230–1232.

Robins, L., Helzer, J., Weissman, M., Orvaschal, H., Gruenberg, E., Burke, J. & Regier D. (1984) Lifetime prevalence of specific psychiatric disorders in three sites. *Archives of General Psychiatry* 41:949–958.

Schlesser, M., Winokur, G. & Sherman, B. (1979) Genetic subtypes of unipolar primary depressive illness distinguished by hypothalamic-pituitary-adrenal axis activity. *Lancet* ii:739–740.

Spitzer, R., Endicott, J. & Robins, E. (1978) Research diagnostic criteria: Rationale and reliability. *Archives of General Psychiatry* 35:773–782.

Tanna, V., Wilson, A., Winokur, G. & Elston, R. (1989) Linkage analysis of pure depressive disease. *Journal of Psychiatry Research* 23:99–107.

Tsuang, M., Winokur, G. & Crowe, R. (1980) Morbility risks of schizophrenia and affective disorders among first-degree relatives of patients with schizophrenia, mania, depression and surgical conditions. *British Journal of Psychiatry* 137:497–504.

Tsuang, M. & Woolson, R. (1977) Mortality in patients with schizophrenia, mania, depression and surgical conditions. *British Journal of Psychiatry* 130:162–166.

von Knorring, A.-L., Cloninger, C., Bohman, M. & Sigvardsson, S. (1983) An adoption study of depressive disorders and substance abuse. *Archives of General Psychiatry* 40:943–950.

Weissman, M. & Myers, J. (1978) Affective disorders in a U.S. urban community. *Archives of General Psychiatry* 35:1304–1311.

Wender, P., Kety, S., Rosenthal, D., Schulsinger, F., Wortmann, J. & Lande, I. (1986) Psychiatry disorders in biological and adopted families of adopted individuals with affective disorders. *Archives of General Psychiatry* 48:923–929.

Winokur, G. (1978) Mania and depression: Family studies and genetics in relation to treatment. In *Psychopharmacology: A Generation of Progress,* ed. M. A. Lipton, A. D. Mascio & K. F. Killam. New York: Raven Press.

Winokur, G. (1983) The validity of familial subtypes in unipolar depression. *McLean Hospital Journal* 8:17–37.

Winokur, G., Behar, D., VanValkenburg, C. & Lowry, M. (1978) Is a familial definition of depression both feasible and valid? *Journal of Nervous & Mental Disease* 166:764–768.

Winokur, G., Cadoret, R., Dorzab, J. & Baker, M. (1971) Depressive disease: A genetic study. *Archives of General Psychiatry* 24:135–144.

Winokur, G., Clayton, P. & Reich, T. (1969) *Manic Depressive Illness*, St. Louis: C. V. Mosby.

Winokur, G., Morrison, J., Clancy, J. & Crowe, R. (1972) The Iowa II. A blind family history comparison of mania, depression and schizophrenia. *Archives of General Psychiatry* 27:462–464.

Winokur, G., Zimmerman, M. & Cadoret, R. (1988) 'Cause the Bible Tells Me So'. *Archives of General Psychiatry* 45:683–684.

Chapter 9

Genetic Studies
of Anxiety Disorders

RAYMOND R. CROWE

Introduction

Genetic studies play many roles in psychiatric research. In the absence of laboratory methods for validating diagnosis and classification, genetic studies provide information on which disorders run true to form in relatives and therefore appear to be biologically distinct syndromes. Genetic studies can also address issues of diagnostic spectrums: which disorders represent borderline or atypical cases of other disorders, and which may be caused by a common genetic diathesis?

Apart from diagnosis, genetic studies are important in identifying disorders that may have a biological basis. This information is helpful in identifying conditions likely to yield answers to biological research. For instance, adoption studies demonstrating a genetic predisposition to schizophrenia were an important precursor of the biological revolution in modern psychiatry.

With the recent impact of recombinant DNA research on psychiatry, it is of more than academic interest to determine which conditions are heritible and how they are inherited. Clinical molecular genetics today is a science of single locus disease, and it is critical that we identify the diseases most likely to be caused by single-locus mutations if the promise of that science is to be fulfilled. Studies of disease transmission patterns within families can provide clues to whether the transmission can be accounted for by single locus genetics or by multifactorial agents.

Family studies examine the degree to which a disorder is familial and the spectrum of disorders found among relatives of patients. Twin studies provide more direct evidence of genetic etiology, as well as further data on the diagnostic spectrum.

175

Adoption studies are uniquely suited to documentation of genetic predisposition by separating the genetic from the familial environment; but unfortunately, none of the anxiety disorders have been studied with this strategy. Studies of transmission patterns help to determine the mode of inheritance. Finally, linkage studies have emerged as the preeminent genetic attack on disease because of their ability to pin down the ultimate genetic lesion. The purpose of this chapter is to update knowledge on the genetics of each of the anxiety disorders as this field enters the molecular genetic era.

Panic Disorder

Family studies of panic disorder can be divided into early studies of anxiety disorder that did not employ operational diagnostic criteria but most likely contained a large number of panic disorders, and more recent studies that used diagnoses similar to current definitions of panic disorder. The reports can be further divided into those based on direct interviews of family members and those based on family histories taken from the proband.

Six studies were published between 1918 and 1951. Among the family history studies, the rates of anxiety disorder in family members were 14% (McInnes, 1937), 25%–42% (Wood, 1941), 15% (Brown, 1942), and 35% (Cohen et al., 1951). All of the studies but the one by Brown examined controls and found higher rates in the patient families, although the difference in the McInnes study was marginal (14% vs 12%). Another way of looking at familiality is by counting the number of probands with an affected family member. Oppenheimer and Rothschild (1918) found 56% of families so affected compared with 35% of the controls; McInnes found the respective rates to be 28% to 18%, and Cohen et al. reported 67% to 12%. Wheeler et al. (1948) interviewed a group of the siblings from the Cohen et al. (1951) study and found 49% affected, compared with 6% of their control siblings.

The first study to use criterion-based diagnoses was the family history study of Noyes et al. (1978). They diagnosed definite or probable anxiety neurosis in 18% of 619 first-degree relatives in the families of 112 patients with that disorder. This rate compared with a 3% rate in similar relatives of medical and surgical control probands. The rate among women relatives of affected probands

was 24% compared with 13% among men. Forty-one of the probands meeting DSM-III criteria for panic disorder were included
in a family study (Crowe et al., 1983). The age-corrected morbidity
risks among first-degree relatives were 17% for definite and 7%
for probable panic disorder, compared with respective rates of
1.8% and 0.5% among control relatives. The sex difference in illness rates was also seen in the interview study: the morbidity risk
was 33% for women relatives of affected probands and 17% for
male relatives. Not only was the rate of panic disorder high
among the relatives, but 61% of the panic disorder probands had
affected relatives compared with 10% of the control probands.

In a follow-up and family study of a clinic sample from Washington University, panic disorder was diagnosed in 9 of 153
interviewed relatives of probands with any primary anxiety disorder (Cloninger et al., 1981). Moran and Andrews (1985) obtained
family histories on the parents and siblings of 60 agoraphobic probands and diagnosed agoraphobia in 12.5%; 18.6% of the females
and 6.7% of the males. Likewise, Hopper et al. (1987) found 11.6%
of the first-degree relatives of 117 panic disorder probands to be
affected by family history; 14.7% of the female and 8.6% of male
relatives.

Up to this point the review has considered the degree to which
panic disorder is familial. An equally important objective is to better define the phenotype being transmitted in the families. This is
a particularly important question if further genetic studies are
being considered because the success of identifying the mode of
inheritance by segregation analysis or of finding the gene by linkage analysis will be dependent upon correctly classifying
relatives as affected or unaffected.

Family studies of panic disorder and agoraphobic probands
have identified a spectrum of disorders that is transmitted with
panic disorder in families (Crowe et al., 1983; Noyes et al., 1986).
Panic attacks that failed to meet the full symptom or attack frequency criteria of DSM-III accounted for a lifetime morbidity risk
of 7.4% among first-degree relatives of panic disorder patients
compared with a 0.5% risk among control relatives. Similarly,
agoraphobia and panic disorder (without agoraphobia) were
found together in families. Among the relatives of agoraphobics,
the risk for panic disorder was 7.0% and for agoraphobia it was
9.4%. Among the relatives of panic disorder probands (without
agoraphobia) the risk for panic disorder was 14.9% and that for

agoraphobia was 1.7%. Note that the combined risk for either disorder is the same in both groups of families—16.4% and 16.6%—suggesting that the familiar predisposition may be expressed either as pure panic disorder or as panic disorder complicated by agoraphobia depending upon the diagnosis of the proband.

The rate of alcoholism appears to be increased in the families of panic disorder probands as well (Noyes et al., 1978; Noyes et al., 1986). For instance, the risk was 12.9% in the relatives of agoraphobics and 6.6% in those of panic disorder probands. On the other hand, these studies failed to find an increased risk of other anxiety disorders or depressive disorders in these families.

Somewhat different findings were obtained from a family study of major depressive disorder in which some of the probands met criteria for panic disorder in addition to major depressive disorder (Leckman et al., 1983). The relatives of these probands were more than twice as likely to be diagnosed as having major depression, panic disorder, phobic disorder, or alcoholism compared with the depressed probands without panic attacks and with normal controls. A separate analysis of children of the same probands replicated the finding (Weissman et al., 1984). Panic disorder or agoraphobia in addition to depression in the proband increased the risk of both depression and anxiety disorder in the children. The findings of these studies were interpreted as suggesting a comorbidity between affective and anxiety disorders.

On the other hand, findings from an NIMH collaborative study on the psychobiology of depression support the distinction between panic and affective disorder (Coryell et al., 1988). Interviewed relatives of probands with panic disorder and secondary depression had an increased rate of panic disorder, but not of affective disorder. These results support a primary-secondary relationship rather than a comorbidity between the two disorders.

In attempting to resolve these apparently discrepant results, it is notable that the studies that have supported the comorbidity hypothesis have been based on probands with affective disorder, and those that have supported the separate disorder hypothesis have been based on probands with primary panic disorder. A critical test of the hypothesis would require the comparison of probands with primary depression and secondary panic disorder to those with primary panic disorder with secondary depression. The comorbidity hypothesis would predict similar familial findings in the two groups, whereas the primary-secondary hypothesis would predict that the proband's primary diagnosis would breed true.

Twins are uniquely suited to reinforce findings from family studies because they allow the examination of a co-twin who is genetically identical to a proband with a disorder of interest. Two twin studies of anxiety disorders have shown a higher concordance in monozygotic than in dizygotic twins. Slater and Shields (1969) found 7 of 17 monozygotic twin pairs to be concordant for anxiety disorders (41%), compared with 1 of 28 dizygotic pairs (4%). In Torgersen's study 9 of 30 monozygotic pairs (31%), and 5 of 56 dizygotic pairs (9%) were concordant (Torgersen, 1978).

Torgersen has reanalyzed his data using DSM-III diagnoses and has provided the only twin data on carefully defined panic disorder (Torgersen, 1983). Thirteen index monozygotic twins had panic disorder and two of their co-twins had panic disorder, two more had panic attacks, and another two had other anxiety disorders. Sixteen index dizygotic twins had panic disorder but none of their co-twins had panic disorder or panic attacks. However, four did have other anxiety disorders. Moreover, in two pairs one twin had panic disorder and the other had agoraphobia. Despite the limited numbers of twins, these data do support the family study findings that panic disorder, panic attacks, and agoraphobia represent a homogeneous genetic entity, which is genetically distinct from other psychiatric disorders.

Torgersen's twins were part of a larger twin study of neuroses, and in a subsequent analysis, the 150 neurotic twins were separated by discriminant analysis into pure anxiety neurosis, mixed anxiety-depression, and pure neurotic depression (Torgersen, 1985). Since two of the most discriminating factors were panic attacks and avoidance of anxiety-provoking situations, the pure anxiety group likely included a large number of panic disorders and agoraphobics. When probandwise twin concordances were examined, only pure anxiety neurosis showed evidence of heritability. Ten of 28 monozygotic co-twins had pure anxiety neurosis compared with 6 of 48 dizygotic co-twins. Moreover, anxiety neurosis and depression showed no evidence of a common genetic predisposition, arguing against a comorbidity hypothesis in these data.

Anxiety and depressive symptoms were studied in the responses to a questionnaire survey of 3,798 twin pairs from the Australian National Health and Medical Research Council Twin Register (Kendler et al., 1986). The study focused on seven anxiety symptoms that included worrying, breathlessness/palpitations, restlessness, panic, head pain, worry/wakefulness, and anxious indecisiveness. The results indicated that genetics accounted for a substantial amount of the variance (33%–46%) in the majority of these common anxiety complaints. The unexpected finding was that familial environment accounted for a trivial amount of the variance, compared with genetic and extrafamilial environmental influences. Although familial environment is often considered to be important in the less severe psychiatric disorders, this did not prove to be the case in this instance. It would be interesting to know which psychiatric diagnoses these symptoms related to in the twins.

If panic disorder is caused by a genetic predisposition as the family and twin findings suggest, the next question to be addressed is how it is inherited. This question is of more than academic interest because modern genetic technology has made it possible to locate genes causing monogenic diseases by searching for linkage to a genetic marker. Diseases fitting a monogenic pattern of inheritance are prime candidates for linkage analysis.

The two genetic hypotheses most commonly considered are the single major locus and the polygenic models. The former considers the disease to be caused by an allele at a single locus with incomplete penetrance. Under the polygenic model a liability to the disease is contributed to by both multiple genes of small, equal, and additive effects and by environmental factors. If a theoretical threshold of liability is exceeded the disease is expressed; otherwise the individual remains unaffected.

An early test of these two hypotheses assumed that familial transmission would be primarily unilateral in the monogenic case and bilateral in the polygenic (Slater, 1966). As implemented, the test compares the number of affected pairs of ancestral relatives that are unilateral (both members on either the paternal or the maternal side) with the number that are bilateral (one member of the pair on each side). Pauls et al. (1979) applied the test to 15 pedigrees of panic disorder and found 37 unilateral to 4 bilateral ancestral pairs, strongly supporting a monogenic hypothesis of transmission.

A more powerful test is provided by segregation analysis, which analyzes transmission on a family-by-family basis and tests the fit to that expected under a genetic model for a given set of parameters. Pauls et al. (1980) analyzed 19 panic disorder pedigrees for fit to a single locus model and found a set of parameters that provided a statistically close fit. The best-fitting parameters predicted an allele frequency of 0.014 and a penetrance of 0.75, with a mean age of onset of 22 years.

One aspect of panic disorder not allowed for by the above models is the sex effect in penetrance. Epidemiological data are in agreement that women are affected two to three times as frequently as men. This can be modeled genetically by assuming that either a single locus or a polygenic diathesis contributes to a liability, with penetrance determined by sex thresholds—for example, women would have a lower threshold, and therefore, would be affected more often than men. This hypothesis was tested with both single-locus and multifactorial sex threshold models, and both models gave a statistically acceptable fit to the observed prevalence rates in families (Crowe et al., 1983). The best-fitting single-locus model predicted an allele frequency of 0.05 and a female and male population prevalence of 4.5% and 2.5%, respectively. The best fitting multifactorial model predicted parent-offspring and sibling-sibling correlations of 0.27 and population prevalences of 17% for women and 9% for men. Since these last two estimates are unrealistic in view of the observed population prevalence rates, the data were reanalyzed constraining the female and male prevalences to 10% and 5%, respectively, and the model still fit the observed data.

In the final analysis, neither competing model could be excluded by the pedigree data, but the fact that single locus genetics can explain the transmission of panic disorder provides some encouragement for pursuing molecular genetic strategies.

The availability of an increasingly large library of DNA probes suitable for use as genetic markers and the imminent prospect of a complete linkage map of the human genome have made genetic linkage a prime strategy for studying the genetics of psychiatric disorders (Martin, 1987). The strategy is to test markers systematically until genetic linkage is detected by means of the disease cosegregating with one allele of the marker. The disease locus is then flanked by a linked marker on the side opposite the first marker, and the DNA between the two markers is cloned and used to develop additional markers successively closer to the disease locus. Once the region has been adequately narrowed, candidate genes can be searched for until the disease locus is discovered. Then by cloning and studying the disease allele in normals and affected individuals some hints as to the function of the gene and the effect of the disease causing mutation can be derived. The ultimate goal is an understanding of the disease pathophysiology through this "reverse genetics" approach.

Linkage between panic disorder and 29 blood type and protein electrophoretic polymorphisms was studied in 198 members of 26 pedigrees (Crowe et al., 1987). Since 39% of these subjects were affected (definite or probable panic disorder or agoraphobia with panic attacks), the pedigrees had reasonable statistical power to detect linkage. Linkage was analyzed assuming an autosomal locus with genetic parameters similar to those estimated by Pauls et al. (1980) but with sex-specific penetrances. The results excluded close linkage to 18 of the marker loci, and revealed a lod score of 2.23 at the alpha-haptoglobin (HP) locus on chromosome 16q22. Since a lod score greater than 3.0 is required for statistical significance, the finding was inconclusive but certainly suggestive of linkage and deserving of further study.

By the time this study was published a DNA probe for the HP locus had become available, so 10 new pedigrees were collected and analyzed for linkage using the same genetic model as before (Crowe et al., 1990). Unfortunately, the new pedigrees excluded close linkage to HP. While the difference between the two studies could represent genetic heterogeneity, it was deemed unlikely since both sets of pedigrees were drawn from the same population and no statistical evidence of heterogeneity of lod scores was found. A parsimonious explanation for the findings is that a Type I statistical error occurred among the 29 analyses performed, and no evidence for linkage on chromosome 16q22 now exists, although this would be an important locus for future linkage studies to examine.

Before leaving the subject of panic disorder some attention needs to be given to its relationship to mitral valve prolapse. The association was first proposed by Wooley (1976) based on the similarity of the clinical pictures described in the literature. Moreover, the epidemiology of the two disorders is similar, with both affecting about 5% of the population, both affecting women twice as frequently as men, and both being familial. At least 12 studies have examined the frequency of echocardiographically diagnosed mitral prolapse in patients with panic disorder, and the rates range from 0% to 50% with the median being 35% (Crowe, 1988). These figures compare with a population prevalence of 5.3% (Savage et al., 1983). Thus, most investigators have found an increased incidence of mitral prolapse in panic disorder patients drawn from clinic populations.

In order to prove an association it would be necessary to demonstrate the association in an unselected population, and this could be examined in either population samples or relatives of affected patients. The best data employing the first approach come from the Framingham population (Savage et al., 1983). There were 208 persons diagnosed as having mitral prolapse by echocardiography and 2,717 with normal echocardiograms. Although a systematic screen for panic disorder was not part of the interview, a number of characteristic symptoms were systematically assessed. These included chest pain, dyspnea, and syncope; and their frequency was the same in both groups.

Psychiatric interviews and family histories were obtained on 50 cardiology clinic patients with echocardiographically diagnosed mitral prolapse, and 12 of the 50 were found to have panic attacks or panic disorder. However, family histories of panic disorder were found almost exclusively in the probands with panic disorder. Since mitral prolapse alone in the proband was not associated with panic disorder in the families this study did not support a familial association between the two disorders. This conclusion is complemented by a family study of 88 patients with echocardiographically diagnosed mitral prolapse (Devereux et al., 1986). Echocardiograms and psychiatric interviews on relatives failed to document an association between mitral prolapse and panic attacks.

These findings make it unlikely that mitral valve prolapse accounts for a subgroup of patients with panic disorder. But if this is the case, how can the association between the two disorders found in clinical populations be explained? One possibility is a selection bias–persons with two diseases are more likely to be found in clinic populations than those with one disease. However, since population studies have shown that mitral prolapse is usually silent, it is difficult to account for such strong associations in psychiatric clinic populations. Another possibility is that panic attacks could predispose the mitral valve to prolapse. This explanation is supported by the observation that mitral prolapse in panic disorder patients often disappears when the panic attacks are successfully treated (Gorman et al., 1981). Finally, it is possible that mitral prolapse might be one of a variety of physical stresses that can precipitate panic attacks in persons who are genetically predisposed. Anecdotally, clinicians often hear that the illness began concomitant with some physical illness and persisted after recovery from the illness. This explanation would be consistent with a strong association in patient populations that would be too weak to be detected in unselected populations.

Generalized Anxiety Disorder

Generalized anxiety disorder is a relative newcomer to the diagnostic nomenclature. Information on its transmission is limited to a single family study (Noyes et al., 1987), in which 123 relatives of 20 probands with generalized anxiety disorder were interviewed and compared with the families of panic disorder patients with and without agoraphobia and with normal controls. The rate of generalized anxiety disorder in the relatives of the probands of that diagnosis was 19.5%, which was significantly higher than the rates found in the other three groups of families. Second, generalized anxiety disorder was the only anxiety disorder that was found more frequently in the families of generalized anxiety disorder probands. The rates of panic disorder, agoraphobia, simple and social phobia, and obsessive-compulsive disorder were no higher than those found among the control relatives. Finally, neither alcohol nor affective disorders were increased in the families of the generalized anxiety disorder probands. These findings support a familial predisposition to generalized anxiety disorder and indicate that the predisposition is specific to generalized anxiety and is not a general liability to any anxiety disorder.

Obsessive-Compulsive Disorder

Although we are witnessing increasing interest in obsessive-compulsive disorder, the literature on this disorder is still limited and it may be a number of years before extensive genetic data are available. Lewis (1935) noted obsessional traits in 37 of 100 parents of obsessional neurotics, as well as in 43 of their 206 siblings. Brown (1942) found obsessional states in 7.5% of parents and 7.1% of siblings of obsessional patients. These rates were higher than those in the families of his other anxiety disorder patients and controls. Rudin (1953), reviewed in Slater and Cowie (1971), reported obsessional illness in 5% of the parents and 2% of the siblings of obsessional patients. Lo (1967) reported pronounced obsessional traits in 8.6% of parents and 4.6% of siblings of obsessional patients. Kringlen (1965) obtained a history of obsessional traits in 18 of 182 (10%) of the patients in his follow-up investigation. On the other hand, Rosenberg (1967) found obsessional neurosis in only 2 first-degree relatives of his probands, or 1%.

As part of a twin study of obsessional illness, Carey and Gottesman (1981) obtained careful family histories of hospitalization or outpatient treatment for obsessional illness (i.e., "those who received psychiatric care and whose obsessions formed a prominent part of their illness notwithstanding other diagnoses"). Using this definition, obsessional illness was diagnosed in 6.2% of the parents, 3.2% of the siblings, and 6.2% of the children. Hospitalization for obsessional illness had occurred in 2.0%, 1.6% and 3.1% of the three classes of relatives, respectively. Although no population rates were available, the hospitalization rates clearly indicate an increased rate of serious obsessional illness among the relatives.

McKeon and Murray (1987) reported a family study of RDC diagnosed obsessive-compulsive clinic patients, with the diagnoses of the family members based on SADS-L interviews supplemented by an obsessional inventory. A control group was included, and although the interviews were not conducted blindly, the diagnoses were. Only one case of obsessive-compulsive disorder was diagnosed among the 149 relatives of the patients (0.7%) and one among the 151 control relatives (0.7%). On the other hand, an excess of psychiatric disorder was found in the relatives of the obsessive-compulsive patients—17% were diagnosed as having depressive, anxiety, phobic, or obsessive-compulsive disorders, compared with 9% of the controls.

Carey and Gottesman (1981) studied 49 twin pairs from the Maudsley Twin Register with hospital diagnoses of obsessional neurosis, obsessional personality, or phobic neurosis. The pairwise monozygotic concordance rate for obsessive symptoms

or features in 15 twin pairs was 87% compared with the dizygotic concordance rate in 15 pairs of 47%. The respective concordance rates for medically treated obsessive symptoms were
33% and 7%.

Some interesting insights into the genetics of obsessive symptoms have come from family studies of Tourette syndrome (Pauls
& Leckman, 1986). First-degree relatives of Tourette syndrome patients were found to have an increased rate of both Tourette
syndrome and chronic motor tics, the respective rates being
10.7% and 18.4% for the combined sexes. Of particular interest
was an increased rate of obsessive-compulsive disorder, affecting
17.2% of the females and 6.7% of the males. This finding suggests
that some cases of obsessive-compulsive disorder may be genetically related to Tourette syndrome.

Phobic Disorders

The familial findings in agoraphobia have already been referred
to with respect to the place of agoraphobia in the panic disorder
spectrum. Several family studies of phobic disorders have been
reported, and most deal with agoraphobia. Buglass et al. (1977)
found a family history of phobic disorder in 7% of the parents and
8% of the siblings of agoraphobic women, compared with 2% and
2%, respectively. Solyom et al. (1974) found a family history of
phobias in 9.5% of the nuclear families and 33% of the extended
families of phobic patients, 91% of which were agoraphobic.
Rates of school phobia as high as 14% have been reported in the
children of agoraphobic women, although no comparison group
of children was included (Berg, 1976). Munjack and Moss (1981)
did not include phobias in their family histories of agoraphobic
patients, but they found an increased rate of alcoholism and
depression in the families. Finally, the study by Noyes et al.
(1986) reviewed above noted a 7.0% rate of panic disorder and a
9.4% rate of agoraphobia in the first-degree relatives of agoraphobics.

Carey and Gottesman (1981) examined pairwise twin concordance rates for phobic symptoms in the Maudsley twin series. The
monozygotic concordance rate for phobic symptoms was 88% in 8
twin pairs and the dizygotic rate was 38% in 13 pairs. The respective rates of medically treated phobias were 13% and 8%.

Conclusion

The strongest case for a genetic predisposition can be made for panic disorder, where the data are consistent with a single locus mutation with greater penetrance in women than in men. These findings recommend panic disorder as a good candidate for linkage studies, and some preliminary work along that line has already begun. Conclusive evidence that panic disorder is a genetic condition would require the finding of linkage indicating the existence of a disease locus.

Progress in biological research in panic disorder has led to increasing interest in other anxiety disorders as well, and studies of obsessive-compulsive disorder, phobic disorders, and generalized anxiety disorders are beginning to appear. It is hoped that the new approaches offered through molecular genetic technology will lead to an understanding of the causes of some of these common and often disabling disorders.

Acknowledgment

Supported by a Research Scientist Development Award from the NIMH #K02MH00735.

References

Berg, I. (1976) School phobia in the children of agoraphobic women. *British Journal of Psychiatry* 128:86–89.

Brown, F. W. (1942) Heredity in the psychoneuroses. *Proceedings of the Royal Society of Medicine* 35:785–790.

Buglass, D., Clarke, J., Henderson, A. S., Kreitman, N. & Presley, A. S. (1977) A study of agoraphobic housewives. *Psychological medicine* 7:73–86.

Carey, G. & Gottesman, I. I. (1981) Twin and family studies of anxiety, phobic, and obsessive disorders. In *Anxiety: New Research and Changing Concepts*, ed. D. F. Klein & J. G. Rabkin, pp. 117–136. New York: Raven Press.

Cloninger, C. R., Martin, R. L., Clayton, P. & Guze, S. B. (1981) A blind follow-up and family study of anxiety neurosis: Preliminary analysis of the St. Louis 500. In *Anxiety: New Research and Changing Concepts*, ed. D. F. Klein & J. G. Rabkin. New York: Raven Press.

Cohen, M. E., Badal, D. W., Kilpatrick, A., Reed, E. W. & White, P. D. (1951) The high familial prevalence of neurocirculatory asthenia (anxiety neurosis, effort syndrome). *American Journal of Human Genetics* 3:126–158.

Coryell, W., Endicott, J., Andreasen, N. C., Keller, M. B., Clayton, P. J., Hirschfeld, R. M. A., et al. (1988) Depression and panic attacks: The significance of overlap as reflected in follow-up and family study data. *American Journal of Psychiatry* 145:293–300.

Crowe, R. R. (1988) Mitral valve prolapse and anxiety. In *Mitral Valve Prolapse and the Mitral Valve Prolapse Syndrome*, ed. H. Boudoulas & C. F. Wooley. Mt. Kisco, NY: Futura Publishing.

Crowe, R. R., Noyes, R., Jr., Pauls, D. L. & Slymen, D. (1983) A family study of panic disorder. *Archives of General Psychiatry* 40:1065–1069.

Crowe, R. R., Noyes, R., Jr., Samuelson, S., Wesner, R. & Wilson, R. (1990) Close linkage between panic disorder and alpha-haptoglobin excluded in 10 families. *Archives of General Psychiatry* 47:377–380.

Crowe, R. R., Noyes, R., Jr., Wilson, A. F., Elston, R. C. & Ward L. J. (1987) A linkage study of panic disorder. *Archives of General Psychiatry* 44:933–937.

Devereux, R. B., Kramer-Fox, R., Browe, W. T., Shear, M. K., Hartman, N., Kligfield, P., et al. (1986) Relation between clinical features of the mitral prolapse syndrome and echocardiographically documented mitral valve prolapse. *Journal of the American College of Cardiology* 8.

Gorman, J. M., Fyer, A. F., Glicklich, J., King, D. L. & Klein D. F. (1981) Mitral valve prolapse and panic disorders: Effect of imipramine. In *Anxiety: New Research and Changing Concepts*, ed. D. F. Klein & J. G. Rabkin. New York: Raven Press.

Hopper, J. L., Judd, F. K., Derrick, P. L. & Burrows, G. D. (1987) A family study of panic disorder. *Genetic Epidemiology* 4:33–41.

Kendler, K. S., Heath, A., Martin, N. G. & Eaves, L. J. (1986) Symptoms of anxiety and depression in a volunteer twin population. *Archives of General Psychiatry* 43:213–221.

Kringlen, E. (1965) Obsessional neurotics: A long-term follow-up. *British Journal of Psychiatry* 111:709–722.

Leckman, J. F., Weissman, M. M., Merikangas, K. R., Pauls, D. L. & Prusoff, B. A. (1983) Panic disorder and major depression: Increased risk of depression, alcoholism, panic, and phobic disorders in families of depressed probands with panic disorder. *Archives of General Psychiatry* 40:1055–1060.

Lewis, A. (1935) Problems of obsessional illness. *Proceedings of the Royal Society of Medicine* 29:325–336.

Lo, W. H. (1967) A follow-up study of obsessional neurotics in Hong Kong Chinese. *British Journal of Psychiatry* 113:823–832.

Martin, J. B. (1987) Molecular genetics: Applications to the clinical neurosciences. *Science* 238:765–772.

McInnes, R. G. (1937) Observations on heredity in neurosis. *Proceedings of the Royal Society of Medicine* 30:23–32.

McKeon, P. & Murray, R. (1987) Familial aspects of obsessive-compulsive neurosis. *British Journal of Psychiatry* 151:528–534.

Moran, C. & Andrews, G. (1985) The familial occurrence of agoraphobia. *British Journal of Psychiatry* 146:262–267.

Munjack, D. J. & Moss, H. B. (1981) Affective disorder and alcoholism in families of agoraphobics. *Archives of General Psychiatry* 38:869–871.

Noyes, R., Jr., Clancy, J., Crowe, R., Hoenk, P. R. & Slymen, D. J. (1978) The familial prevalence of anxiety neurosis. *Archives of General Psychiatry* 35:1057–1059.

Noyes, R., Jr., Clarkson, C., Crowe, R. R., Yates, W. R. & McChesney, C. M. (1987) A family study of generalized anxiety disorder. *American Journal of Psychiatry* 144:1019–1024.

Noyes, R., Jr., Crowe, R. R., Harris, E. L., Hamra, B. J., McChesney, C. M. & Chaudhry, D. R. (1986) Relationship between panic disorder and agoraphobia: A family study. *Archives of General Psychiatry* 43:227–232.

Oppenheimer, B. S. & Rothschild, M. A. (1918) The psychoneurotic factor in the irritable heart in soldiers. *Journal of the American Medical Association* 70:1919–1923.

Pauls, D. L., Bucher, K. D., Crowe, R. R. & Noyes, R. (1980) A genetic study of panic disorder pedigrees. *American Journal of Human Genetics* 32:639–644.

Pauls, D. L., Crowe, R. R. & Noyes, R. (1979) Distribution of ancestral secondary cases in anxiety neurosis (panic disorder). *Journal of Affective Disorders* 1:287–290.

Pauls, D. L. & Leckman, J. F. (1986) The inheritance of Gilles De La Tourette's syndrome and associated behaviors: Evidence for autosomal dominant transmission. *New England Journal of Medicine* 315:993–997.

Rosenberg, C. M. (1967) Familial aspects of obsessional neurosis. *British Journal of Psychiatry* 113:405–413.

Rudin, E. (1953) Ein Beitrag zur Frage der Zwangskrankheit, insbesondere ihrer hereditaren Beziehungen. *Archiv fur Psychiatrie und Zeitschrift Neurologie* 191:14–54.

Savage, D. D., Garrison, R. J., Devereux, R. B., Castelli, W. P., Anderson, S. J., Levy, D., et al. (1983) Mitral valve prolapse in the general population. I. Epidemiological features: The Framingham Study. *American Heart Journal* 106:571–576.

Slater, E. (1966) Expectation of abnormality on paternal and maternal sides: A computational model. *Journal of Medical Genetics* 3:159–161.

Slater, E. & Cowie, V. (1971) *The Genetics of Mental Disorders*. London: Oxford University Press, 105–106.

Slater, E. & Shields, J. (1969) Genetical aspects of anxiety. *British Journal of Psychiatry* 3 (special publication):62–71.

Solyom, L., Beck, P., Solyom, C. & Hugel, R. (1974) Some etiological factors in phobic neurosis. *Canadian Psychiatric Association Journal* 19:69–78.

Torgersen, S. (1978) Contribution of twins to psychiatric nosology. In *Twin Research: Part A. Psychology and Methodology* ed. W. E. Nance. New York: Alan R. Liss.

Torgersen, S. (1983) Genetic factors in anxiety disorders. *Archives of General Psychiatry* 40:1085–1089.

Torgersen, S. (1985) Heredity differentiation of anxiety and affective neuroses. *British Journal of Psychiatry* 146:530–534.

Weissman, M. M., Leckman, J. F., Merikangas, K. R., Gammon, G. D. & Prusoff, B. A. (1984) Depression and anxiety disorders in parents and children: Results from the Yale Family Study. *Archives of General Psychiatry* 41:845–852.

Wheeler, E. O., White, P. D., Reed, E. & Cohen, M. E. (1948) Familial incidence of neurocirculatory asthenia ("anxiety neurosis," "effort syndrome"). *Journal of Clinical Investigation* 27:562.

Wood, P. (1941) Aetiology of Da Costa's syndrome. *British Medical Journal* 1:845–851.

Wooley, C. F. (1976) Where are the diseases of yesteryear? DaCosta's syndrome, soldier's heart, the effort syndrome, neurocirculatory asthenia, and the mitral valve prolapse syndrome. *Circulation* 53:749–751.

Chapter 10

The Contribution
of Heredity to Obsessional
Disorder and Personality:
A Review of Family
and Twin Study
Evidence

ALISON M. MACDONALD
ROBIN M. MURRAY
CHRISTINE A. CLIFFORD

Obsessive-compulsive neurosis is a fascinating and yet frustrating disorder for researchers. The reasons are twofold: first, the difficulty in reliably defining and assessing the symptoms, and second, the rarity of the complaint as presented at psychiatric assessment. Studies described elsewhere in this volume show that there is a considerable familial, and therefore possibly genetic, influence on the transmission of anxiety disorders, with which obsessive-compulsive disorder is traditionally grouped in diagnostic classifications. There also appears to be considerable genetic influence on measures of neurotic symptoms in the general population and on the Eysenckian personality dimension of neuroticism (Kendler et al., 1987). However, studies of obsessive-compulsive disorder (OCD) itself remain far from conclusive. It was not without some justification that Lewis (1936:325) asserted that "it may well be that obsessional illness cannot be understood altogether without understanding the nature of man."

Obsessional Disorders: Definition
and Epidemiology

Obsessions are traditionally defined as intrusive, repetitive, and unwanted thoughts or impulses, which the sufferer recognizes as his or her own, but usually tries to resist (Rachman & Hodgson, 1980). As Lewis (1957) put it, "The essence of an obsession is the fruitless struggle against a disturbance." Obsessions may be accompanied by repetitive stereotyped patterns of behavior termed compulsions.

Only a small proportion of those who seek psychiatric help do so because of obsessions. Hare et al. (1972) calculated that obsessional disorders accounted for only 3% of all neuroses and 0.5% of all admissions to psychiatric hospitals in England and Wales. There is some evidence that both obsessionality and obsessional neurosis are more common in Ireland than elsewhere in Britain (Scott et al., 1982). Series of obsessional neurotics have been reported not only from the industrialized West but also from China, India, and the Arabic countries. There is, however, no substantial information about the comparative frequency of obsessional neurosis in different cultures.

Obsessional neurosis most frequently begins in late childhood and early adult life; 50% of cases have their onset before age 20 and only 5% after 40 years of age. Until recently the prognosis was poor. Kringlen (1965) believed that most obsessionals had a miserable life, while Rachman and Hodgson (1980:2–3) found it hard to imagine "the amount of suffering involved, and how it can reach such proportions as to imprison the person and prevent him or her carrying out any constructive work."

The study of the clinical syndrome of obsessive-compulsive disorder from a genetic viewpoint has always been difficult because of its relative rarity. The oft-quoted 0.05% (Rudin, 1953; Woodruff & Pitts, 1964) maximum prevalence rate of OCD in the general population is an estimate based on the combination of a 0.5% rate in a psychiatric population and a 1% rate of psychiatric attendance in the general population. In those earlier epidemiological studies of neurosis which broke down the categories and gave figures for prevalence of OCD, there is a wide variation from less than 0.01% (Bille & Juel-Nielsen, 1968; Lemkau et al., 1942) to 2.5% (Vaisanen, 1976). These differences may be partly accounted for by study methods ranging from using record data only to employing personal interviews. One study of the prevalence of neurosis that used both these methods found a tenfold difference in prevalence of all neurosis (15.7% from interview versus 1.6% from records). Brunetti (1976 cited by Carey et al., 1980) reported that 0.98% of personally interviewed subjects were diagnosed OCD compared with 0.14% from records.

The recent Epidemiological Catchment Area (ECA) studies in the United States (Karno et al., 1988), using the structured Diagnostic Interview Schedule (DIS; with more than 18,000 people, suggest that the lifetime prevalence rates in the general population may be much higher than previously thought. This has led to recent references to a "hidden epidemic" (Jenike, 1989). Using DSM-III criteria, lifetime prevalence for OCD ranged from 1.9% to 3.3% (in different study areas) when DSM-III exclusions were not made (presence of Gilles de la Tourette's syndrome, schizophrenia, major depressive disorder, organic mental disorder) and from 1.2% to 2.4% when schizophrenia and major depressive disorder were specifically excluded.

This study cannot be regarded as the last word on the subject, as there are considerable problems with the use of lay interviewers to detect clinical OCD; indeed Karno et al. point to the extremely low agreements between the DIS rating and physician diagnoses of the same patients, with kappas as low as 0.05, for example, only chance agreement. The study found that only 8.6% of supposedly OCD individuals reported both obsessions and compulsions, in marked contrast with a study of 150 inpatient obsessionals which found that 69% suffered both obsessions and compulsions (Welner et al., 1976, cited by Rachman & Hodgson, 1980:12). It may be, therefore, that in the general population, individual symptoms and occasional disturbance occur much more commonly than the clinical syndrome seen by therapists. Such a finding suggests a multiple threshold approach for genetic studies of relatives, for example, retaining the categorical approach to illness data but postulating a continuous underlying liability distribution with OCD being manifest when a certain level of liability is reached, and perhaps obsessional personality representing a lower threshold. An alternative view is of a dimensional exophenotype, with clinical cases being regarded as the extreme end of a continuum on which most of the population have only the odd symptom.

Family Studies

It is usually accepted that if a disorder clusters in the relatives of index cases, there is likely to be some familial, and therefore possibly some genetic, factor in the etiology. This idea is tempered by the fact that apparently high rates in relatives must be compared with population prevalence rates or an appropriately large control group, and that the criteria for diagnosis in relatives must be comparable with those for the index cases. Also, of course, familial does not necessarily mean genetic; correlation of liability is frequently misunderstood as a genetic concept, although it represents all sources of familial resemblance unless explicit assumptions are made or data are available which allow genetic and environmental sources of resemblance to be partitioned (Reich et al., 1972).

A number of investigators have studied the relatives of obsessional patients with regard to rates of OCD, as well as rates of obsessional traits and personality that might be regarded as milder forms of expression of a common underlying liability. Results of these studies are summarized in Table 10.1. Many of the finer details of the earlier studies have been discussed extensively by other reviewers (see for example, Carey & Gottesman, 1981; Rachman & Hodgson, 1980), so we will focus on the extent to which the rates observed in relatives allow us to draw conclusions about the genetic contributions to OCD.

There are methodological deficits in most of the studies, particularly the earlier ones; information on relatives was often not first-hand, and the relatives were diagnosed with knowledge of the proband's diagnosis. However, if one takes into account the different sampling techniques and the varying nosological criteria used, then there is a surprising amount of agreement concerning obsessional neurosis, which appears to affect between 4.6% and 10% of parents in most earlier, and in two out of three recent, studies. There is much less agreement over the frequency of obsessional personality, with rates for parents ranging from 3.3% (Rudin, 1953) to 37% (Lewis, 1936). This presumably reflects different concepts of obsessional personality in the various investigations, as well as the unreliability of psychiatrists' categorical descriptions of personality types (Walton & Presly, 1973). A high proportion of relatives appear to have some psychological abnormality; for example, Carey (1978) found that 48% of parents, 39% of siblings, and 16% of children had some form of noteworthy abnormality. However, none of the studies noted any increased liability to psychosis.

One other family study (Alanen et al., 1966), which included just five OCD index cases, found that about 40% of the first-degree relatives had some obsessional features included in a

TABLE 10.1.
Family Studies of Obsessive-Compulsive Disorder

Study	No of index cases	Controls (N)	Type of relatives	Rates in Relatives. % (Control Relatives)		
				Obsessive-compulsive disorder	Obsessional traits/personality	Any abnormality
Luxenburger (1930)**	71		parents sibs	10 6.0	8.0	66.0
Lewis 1936	50	none	parents sibs		37.0 20.9	56.0* 45.6
Brown (1942)	20	medical patients (31)	parents sibs	7.5 (0) 7.1 (0)		50.0 (19.4) 35.7 (10.6)
Rudin (1953)	130		parents sibs	4.6 2.3	3.3 3.1	40.2 19.1
Kringlen (1965)	91	other neurotics (91)	parents		9.9	47.8
Lo (1967)	88	none	parents sibs		8.6 4.6	19.4* 10.7
Rosenberg (1967)	144	none	parents sibs	0.4–1.2 (hospitalized)		9.3–10.9 (treated only)
Carey (1978)**	24 twin pairs	none	parents sibs	2.0–6.2 1.6–3.2	8.3–27.0 6.4–19.3	47.9 38.7
Rasmussen & Tsuang (1986)	44	none	parents	5	11	44.3*
McKeon & Murray (1987)	50	GP attenders (50)	parents sibs	0.7 (0.7)		27.5 (13.9)

*Rates are maximum, obtained by summing reported diagnoses in relatives.
**As cited by Carey and Gottesman (1981).

diagnostic formulation, and closer to 70% had some abnormality. Rosenberg (1967) found a much lower rate than other studies, perhaps because he included only hospital-treated OCD among relatives as affected. Rosenberg did not report rates of obsessional personality or other symptoms in relatives.

Thus the rates of OCD and obsessionality seem to be generally high in relatives of index cases when these figures are compared with the various published prevalence rates. In most cases where controls were used, allowing more appropriate comparisons, the rates of OCD were higher in relatives of index cases, although this was not the case in the most recent study (McKeon & Murray, 1987), in which just one OCD relative was found in either group of relatives, giving a rate of 0.7%. In this, the only study in which relatives have been systematically examined for other operationally defined diagnoses, the relatives of OCD cases were found to have high cases of neurotic disorders in general. McKeon and Murray (1987) found a statistically significant excess of SADS-L interview diagnoses among index relatives compared with control relatives, largely accounted for by anxious, phobic, and depressive disorders.

Rasmussen and Tsuang (1986) studied 44 DSM-III OCD patients and provided data on the prevalence in parents based on reports given by the patients. As with other studies which have used operationalized criteria, they found that OCD in the index cases frequently coexisted with another anxiety disorder or major depressive disorder. Although 5% of parents were found to have OCD and 11% to have significant obsessional traits, 13.6% met criteria for major depressive disorder, 5.7% for other anxiety disorders, and 9.1% for alcohol abuse. There were no cases of psychosis. The lack of a control group and the use of family history criteria make it impossible to assess the significance of these rates with any confidence.

The problem remains that without reliable and comparable prevalence rates for OCD and obsessional personality in the general population or in a well-chosen large control sample, the figures obtained from assessment of relatives cannot be related to any particular genetic model. Carey et al. (1980) have shown how the widely varying assessments of prevalence rates for neurosis in the general population have dramatic effects on the results of quantitative analysis of familiality. For the relatively rare OCD, with prevalences differing between 0.01% and 2.5% in a range of studies examined, analysis using the 2.3% rate in siblings found in one family study (Rudin, 1953) led to an estimation of sibling correlation for liability to OCD of between -0.02 and 0.51, with correspondingly wide-ranging estimates of heritability. In other words, depending on the prevalence rate used in comparison with the observed rate in relatives, OCD may appear to be either totally nonfamilial or entirely familial.

A number of other recent studies have also examined relatives of OCD cases, often in relation to overlap with affective disorders. Rapoport et al. (1981) collected a wide range of data on 9 adolescents with primary OCD, but who all met DSM-III criteria for major depressive disorder at some point in their lives after the onset of OCD; 1 sibling had OCD (i.e., 1 first-degree relative out of 45 studied, 18 of whom were under age 17). A variety of other disorders occurred in the families, most notably alcoholism in parents and anxiety disorder among the siblings. Only 3 of the 9 siblings over age 18 had no diagnosis at all, the diagnoses of the other 6 all being types of neurosis or substance abuse.

Coryell (1981) applied Feighner Criteria to the records of 110 patients with a hospital diagnosis of OCD (1.1% of all inpatient admissions in the period of study), and found 44 patients who met these criteria (0.45% of the inpatient admissions). Sadly, he did not specifically report rates of obsessional diagnoses, symptoms, or personality in relatives, but only noted that fewer of the relatives of OCD cases had affective disorder than relatives of affective index cases (6.1% versus 21.4%). This is perhaps less than surprising as all comorbid index cases were excluded.

Another family history study (Insel et al., 1983) examined 27 OCD index cases diagnosed by DSM-III criteria. Family history data were collected and some parents were interviewed. None of the 27 cases had a parent with treated OCD or a history of symptoms although 2 were found to show some obsessional behaviors. One patient had a son with checking compulsions, while 3 mothers had a history of treated affective disorder and 3 fathers a history of treated alcoholism.

In these three studies, we again see the occurrence of neurotic disorders other than the rare OCD in the relatives, with depression, other anxiety disorders, and alcohol abuse appearing frequently.

On the basis of an examination of 11 pedigrees through propositae with Gilles de la Tourette's syndrome (TS), Comings and

Comings (1987) suggested that agoraphobia and obsessive-compulsive behavior may be subdivided into categories: (a) sporadic (no family history), (b) familial (positive family history), and (c) cases possibly due to a TS gene (positive family history of motor or vocal tics in index case or family members). In other words, they suggest that a proportion of agoraphobic or OCD patients may be attributed to partial expression of a TS gene. The actual number of family members diagnosed as OCD is not reported so the conclusions seem to be based on the finding that 9 out of 90 of the female relatives were diagnosed agoraphobic with panic, and anecdotal reporting of a high rate of OCD in relatives in association with motor or vocal tics. The link of some cases of OCD with Tourette syndrome has been supported by some (Pauls, 1989) and dismissed as an artifact by others (Shapiro & Shapiro, 1989) but much further evidence is needed before the sweeping conclusions of this paper are warranted.

Several studies have been directed to the children of obsessional patients, in spite of the fact that obsessional patients are less likely to marry than the general population, and within marriage their fertility is low (Hare et al., 1972). Cowie (1961) studied 30 offspring of 20 obsessionals and noted "a higher incidence of neurosis among the offspring of obsessionals than amongst the offspring of parents of other diagnostic categories investigated." Neurotic disturbance was particularly common when the mother was obsessional; 7 children were neurotic and in 6 of those cases it was the mother who had been obsessional. Rutter (1966) investigated 9 children of obsessional parents and again found that they showed more psychological dysfunction than the children of other psychiatrically ill parents. Four had distinct obsessional symptoms.

It is difficult to draw firm conclusions, even from the more recent family studies (despite the improvements in comparability achieved by use of structured interviews and operationalized criteria), because, like the studies conducted several decades ago, they are small. Also, the restriction of measurement to categorization by diagnoses is relatively uninformative, particularly if there are no provisions for grouping subclinical or milder "cases" which might represent a lower threshold of liability. Those studies which attempted measurement of symptoms and traits and other syndromes in the index cases go some way toward providing more information about the relationship of the diagnosis to what may actually be transmitted. However, we are left to think that this may be a nonspecific vulnerability to neurosis in one form or another, with really very little evidence supporting the alternative view of specific transmission of OCD. The rarity of clinical OCD and the variable published prevalence rates make it impossible to say whether the low rates of OCD found in relatives indicate that it is really not partly inherited (genetically or culturally) or simply that the studies are too small to allow proper statistical comparison. Most noteworthy are the rates of obsessional symptoms not amounting to clinical OCD which are found in relatives (Carey, 1978; Rasmussen & Tsuang, 1986) and the apparent raised risk of other neurotic disorders (McKeon & Murray, 1987).

Twin Studies of Obsessive-Compulsive Disorder

Adoption and twin studies are the usual ways of teasing apart the effects of heredity and environment. The methods involve comparison of biologically related individuals with different proportions of shared genes, such as monozygotic and dizygotic twins or singleton sibling pairs reared together or in adoptive families (adopted singletons and reared-apart twins). Adoption studies permit exclusion of shared environmental factors from the correlations in liability and a more accurate estimation of heritability. Although adoption studies have not proved feasible in obsessional neurosis, twins have been examined for this disorder.

A number of authors of twin case reports (Marks et al., 1969; Woodruff & Pitts, 1964) have calculated that the chance of finding a pair of monozygotic (MZ) adult twins, both of whom have OCD, would be between one in 600 million and one in 800 million if the disorder arose independently in each co-twin, and was not due to some combination of shared genetic or environmental factors; the fallacious conclusion has been drawn by some that the finding of even one or two concordant pairs of MZ twins is strong support for a genetic contribution. Using Karno et al.'s (1988) maximum prevalence rate for OCD increases the probability of finding MZ twins concordant by chance to around one in 200,000, somewhat less impressive.

Rachman and Hodgson (1980:41) have criticized the use of these probabilities on the basis that concordance is not "only random," but in such a rare disorder bias will inevitably occur in the "selection, assessment and diagnosis of co-twins of monozygotic obsessional patients." Other reviewers (Black, 1974; Emmelkamp, 1983) have not considered the evidence to be sufficient, or sufficiently good, to assess the contribution of genetic factors to OCD in MZ twins. We would add that for rare OCD, the finding of concordant pairs of twins would be very unusual in the absence of environmental factors contributing to similarity, unless the heritability of the disorder was very high (Smith, 1970).

Carey (1978) estimated that 30 concordant and 13 discordant MZ pairs, and no concordant but 14 discordant dizygotic (DZ) pairs had been reported in the world literature. Unfortunately, one cannot take these figures at their face value. First, most report have concerned only a few cases, and it is well known that twins collected in an unsystematized way tend toward the MZ and concordant. Second, in many cases, both the diagnosis and zygosity are in doubt.

A considerable number of the pairs in the literature are from a series of consecutive admissions to the Maudsley Hospital, London, where records on twins have been maintained systematically for more than 40 years (see, e.g., Gottesman & Shields, 1972; for a description). As data on these twins have been presented in a number of publications, often including some of the same pairs, we will add a separate section on this series to try and avoid furthering the illusion that there are more OCD twins in the literature than really exist.

If we look at the largest non-Maudsley series, we find that Tienari (1963) claimed a 91% (pairwise) concordance rate for 11 MZ pairs, but Black (1974:23) wrote, "There is no support for the diagnosis of obsessional neurosis in a single case." Similarly, Inouye (1965) reported an 80% concordance rate for 10 MZ twins, but Rachman and Hodgson (1980) doubt whether any of the cases were, in fact, obsessive compulsives. Inouye was reporting an English-language version of a study by Ihda (1960) of 25 pairs of neurotic twins; as these were said to comprise 20 MZ and just 5 DZ pairs, and 10 of them were obtained through personal referrals, the biases in ascertainment and doubts about diagnosis and zygosity make this study too flawed to be useful.

More recently, Tarsh (1978) has reported a concordant opposite sex DZ pair. Hoaken and Schnurr (1980) described a discordant MZ pair (although zygosity is said to be MZ on the basis of 10 identical blood groups, and hence likely, no probability of mono-zygosity or details of the blood groups are given). The twins (reported at age 20) had certainly not passed through the age of risk, and cannot therefore be described as definitely discordant. Furthermore, the proband's diagnosis is questionable in the light of possible organic complications, and a lack of resistance to com-pulsions which were largely ego-syntonic.

Rosenberg (1967) mentioned 2 opposite-sexed twins among his index cases; one co-twin was depressed, the other normal. Other reports of OCD in twins included in Carey's (1978) headcount are reported by Braconi (1970) and by Schepank (1976), but in these reports the diagnoses are questionable.

Torgersen (1983) investigated 3 MZ and 9 DZ twin pairs from the Norwegian twin registry with at least one twin having OCD, and found none of them to be concordant for OCD. One mono-zygotic co-twin was diagnosed agoraphobic (without panic) and one dizygotic co-twin met criteria for generalized anxiety disor-der. Three other co-twins had nonanxiety diagnoses which were not specified. Rates of milder obsessional symptoms or person-ality were not reported.

In two further recent case reports of monozygotic twins reported to be concordant for OCD, both have blood grouping diagnoses of zygosity. The male pair (Mahgoub et al., 1988) are Saudi twins with a first-born twin (with a history of grand mal epilepsy) with obsessional doubts and washing rituals. The second-born twin seems to have fairly minor doubts and rituals, in addition to fears of disasters befalling his twin, which do not seem to have interfered significantly with his functioning or caused much distress. A female pair (Kim et al., 1990) are concor-dant for checking rituals, with onset at age 11, resulting in signif-icant distress and causing both to leave their jobs.

The Maudsley Hospital OCD Twins

Many of the twin case reports and studies of OCD to be found in the literature are of twins included in the consecutively ascer-tained Maudsley Hospital Twin Register, collected since 1948 and with reliable zygosity diagnosis through blood grouping in most cases (Carey, 1978; Carey & Gottesman, 1981; Lewis et al., 1991, in press; Macdonald & Murray, 1989; Marks et al., 1969; McGuffin & Mawson, 1980; Parker, 1964a, 1964b). Black (1974) concluded that, of all the cases in the literature at that time, in only three pairs had both zygosity and diagnosis been firmly established (Marks et al., 1969; Parker, 1964a; Woodruff & Pitts, 1964). In the first of these pairs, both twins had evidence of brain damage. The other one concordant and one discordant MZ pairs were both ascertained through index twins attending the Maudsley Hospital. The concordant pair was said by Rachman and Hodgson (1980) to offer more support for a social transmission model than a genetic one, and have subsequently become discordant for schizophrenia (Lewis et al., 1991, in press).

Previous reviews of twin studies have entirely overlooked the fact that Parker (1964b) actually reported not only a discordant MZ pair (for OCD specifically, and separately reported in Parker, 1964a) but also a concordant DZ pair from the Maudsley records. Examination of the records has not enabled us to identify this pair of twins, who may have been a personal referral outside of the consecutive series.

McGuffin and Mawson (1980) reported two concordant MZ pairs ascertained through treatment of the index case at the Maudsley Hospital, one pair of whom (the "W" twins) have been more recently followed up by one of us (AMM). The co-twin now has far more serious OCD problems; the similarity with her previously severely affected sister, who is now functioning well, is striking alongside the proband's reports that she mostly kept the details of her obsessions to herself and did not discuss them with her co-twin.

Carey and Gottesman (1981) reported 15 pairs of MZ twins and 15 pairs of DZ twins, ascertained through consecutive admissions of index cases to the Maudsley Hospital, and including some pairs previously reported (Marks et al., 1969; Parker, 1964a), as well as those examined in more detail in Carey (1978). Thirty-three percent of MZ co-twins had psychiatric or general practitioner treatment involving obsessional symptoms versus 7% of DZ co-twins. Obsessional symptoms with or without treatment were observed in 87% of MZ co-twins and 47% of DZ co-twins. Unfortunately, no further analyses of these carefully collected data (and reliable diagnosis of zygosity), which include questionnaire measures of obsessionality and other traits, have been published to our knowledge. A follow-up of most twins from this series and additional pairs ascertained in the subsequent ten years is in progress; preliminary results (Macdonald & Murray,

1989) largely confirm earlier findings of increasing MZ and DZ concordance as inclusion criteria are broadened towards obsessional traits or any neurotic disorder. This would be expected purely on the basis of higher prevalence of such traits in the population, but the probandwise concordance rates for hospital treated OCD are 45.5% in MZ pairs versus 10.5% in DZ pairs. As has been shown previously, the choice of population rates can have a dramatic effect on estimates of heritability. By using 0.05% (Woodruff & Pitts, 1964) or 1.9% (Karno et al., 1988, lower rate without DSM-III exclusions), the heritability of liability is calculated as 0.52 or 0.90, respectively. We stress that these figures are preliminary, given only because of a lack of other available estimates and to illustrate the problems involved (hence the lack of any standard errors).

Lewis et al. (1991, in press) described three pairs of monozygotic twins ascertained through the Maudsley Twin Register through an index case with either OCD (1) or schizophrenia (2). These twins were all found to be concordant for DSM-III-R OCD but discordant for schizophrenia or schizoaffective disorder. The nonpsychotic co-twins met criteria for schizotypal personality disorder. Only one of these pairs has previously been noted in the literature on OCD (Marks et al., 1969); the other two were ascertained though an index case with a primary diagnosis of schizophrenia, but they are interesting in that not only are they concordant for OCD but they are discordant for the more common diagnosis of schizophrenia.

As with the family studies, it is necessary for twin studies to have either reliable and comparable population prevalence rates or large control groups. In none of the twin studies have controls been used, unless one includes the neurotic twins in Torgersen's study, in whom the only co-twin diagnosed OCD was from a double proband pair in which the other was agoraphobic. The rarity of twins with OCD is a two-edged sword. On the one hand, it is well nigh impossible to design a study grounded in genetic theory and testable models. On the other, the fact that such twins do sometimes occur, however beset with problems, makes them intriguing from a phenomenological viewpoint. They are also of potential value for co-twin control methodology, for example, where other features and environmental factors are studied in discordant MZ twins, to see if such factors influence discordance and hence may be involved in etiology. To our knowledge, this approach has not been explored in MZ discordant obsessionals. On their own, as case studies, it is very easy to point to the variety of environmental factors that could lead to concordance just as readily as shared genes. Apart from systematic long-term studies of substantial series of hospital ascertained twins, the twin method is probably of more value when applied to obsessional traits and symptoms in larger samples of "normal" twins.

Obsessional Neurosis and "Normal" Obsessionality

It has long been recognized that obsessional neurosis may arise in the context of a personality with strong obsessional features. A famous example is John Bunyan who termed his repetitive blasphemous thoughts "pollution of the mind"; Samuel Johnson and Charles Darwin also had strong obsessional traits, and at one time or another all three suffered from obsessional illness. As Rachman and Hodgson (1980:25) point out: "In each case their work bears the marks of their obsessional tendencies—collecting, systematically ordering and organising, making increasingly fine distinctions, struggling with unacceptable thoughts, and so on." The billionaire Howard Hughes is a more recent example. He was haunted by fears of contamination for many years, and his later life was dominated by increasingly bizarre attempts to avoid touching any person or object directly. His biographers, Bartlett and Steele (1979), demonstrate that his obsessionality was one of the sources of his earlier business success, which might lead us to conclude that genetic research, in the form of the Howard Hughes Institutes, has also gained much from obsessionality!

If obsessional neurosis, as described above, is regarded as a distinct disease entity, qualitatively different from normal behavior, then its rarity means that it is very difficult to devise ways of successfully examining any possible etiological role of heredity. In addition, the usefulness of such studies is questionable if the general population has a higher rate of a less severe form of the problem, as suggested by the ECA studies. The contemporary view of neuroses, in Britain at least (Tyrer, 1985), is to consider them as clusters of symptoms to which individuals toward the extreme ends of normally distributed traits or personality dimensions are especially prone, perhaps when exposed to particular life events or patterns of upbringing, or in the context of a particular genetic background (Kendler et al., 1987).

Such a view holds out the prospect that the study of the genetic contribution to obsessionality in normal population samples may produce findings with important implications for the pathogenesis of obsessional disorders. Therefore, we will now discuss obsessional neurosis and its relation to obsessional traits and behavior in the normal population, before going on to review the evidence concerning a possible genetic contribution to obsessionality.

How to describe personality has been a matter of some controversy. Lewis (1938) described the obsessional personality as demonstrating "excessive cleanliness, orderliness, pedantry, conscientiousness, inclusive ways of thinking and acting; perhaps also a fondness for collecting things including money." Black (1974), reviewing retrospective studies, found that premorbid obsessional traits had been noted in 71% of OCD patients. Emmelkamp (1983) reexamined studies of premorbid personality traits and concluded that while the data indicate a strong relationship between premorbid obsessional personality traits and the development of OCD, there is no one-to-one relationship; furthermore, the high rates of obsessional traits in non-OCD patients and in neurotic patients in general, as well as a more general association of OCD with introverted neurotic traits, temper any conclusions which may be drawn about specific causal relationships. As Mapother and Lewis (1941) pointed out, "obsessional traits occur, however, in many people who never become mentally ill, and in many who become mentally ill otherwise than with an obsessional disorder."

Eysenck and Eysenck (1964) conceived of personality in terms of two orthogonal dimensions: (1) introversion—extraversion and (2) stability—neuroticism. They hypothesized that those individuals high on neuroticism would be prone to psychiatric conditions. Those with, in addition, high extraversion would tend to have hysterical disorders, whereas those with low extraversion they considered especially at risk for anxiety states, phobic, and obsessional disorders. The Eysenckian model, which is now so widespread that it has been considered to have achieved the status of a paradigm (Eaves et al., 1989), has provided the basis for many quantitative genetic studies, using twin data in particular.

Although much evidence supports the notion that obsessional neurosis represents one of the extremes of human variation rather than a discrete disease entity quite unrelated to normal patterns of personality and behavior (Cooper & Kelleher, 1973; Rachman & De Silva, 1978), it is true that the majority of those who consult psychiatrists for primary obsessional complaints are categorized as suffering from OCD. However, this categorical classification results from traditional diagnostic practices; it does not alter the fact that most British psychiatrists and psychologists now believe obsessional neurotics differ from normal subjects quantitatively rather than qualitatively. Indeed, it is difficult to know where normality ends and abnormality begins. At what point does handwashing become excessive—ten, twenty, or thirty times daily? When does commendable tidiness become pathological overtidiness? There can be no definite answers to these questions, and we must be careful not to confuse statistical abnormality with high levels of subjective distress and actual impairment, which lead to treatment seeking.

Twin Studies of Obsessionality as a Dimension

The study of continuously distributed traits, such as obsessional trait or symptom scores in twins, allows the use of different analyses from those which are available for categorical data like diagnoses. It also removes the need for population prevalence rates for comparison. Instead of relating diagnoses to thresholds on a hypothesized liability distibution, we may hypothesize that our measures are more direct indices of liability and hence examine the genetic and environmental basis of individual differences in vulnerability to develop clinically significant OCD (which in this case is viewed as the equivalent of extreme scores on trait/symptom measures). Of course, such an approach would be useful if applied to relevant behavioral measures, or physiological data. In the study of neurosis in general, twin studies of the relationship between symptoms of anxiety and depression and the neuroticism personality dimension have appeared in the last few years using ever more sophisticated techniques (Eaves et al., 1989). For obsessionality specifically, they are limited.

Young et al. (1971) conducted a small study of 32 pairs of male twins who completed the Middlesex Hospital Questionnaire, which contains a brief obsessional traits and symptoms subscale. They found no significant difference between MZ and DZ intraclass correlations for obsessional subscale scores, unlike the various anxiety subscales and neuroticism correlations, which were significantly different, suggesting a genetic influence on these but not on obsessionality.

Torgersen (1980) examined 99 pairs of same-sex twins from the Norwegian register, including 11 pairs selected because one twin had been hospitalized for neurotic problems. Using a questionnaire derived from the psychoanalytic literature by Lazare et al. (1970), previously used only on patient samples, Torgersen substantially replicated the factor structure in the "normal" twin group on oral, obsessive, and hysterical factors. He broke down the figures for MZ and DZ twins, and separated the sexes, so there were less than 30 pairs in each group.

Torgersen compared the intraclass correlation coefficients, and intrapair variances, for MZ and DZ twin pairs, and found no evidence suggesting that the obsessionality factor was influenced by heredity. However, it is difficult to evaluate the significance of the findings. Torgersen reported that some of the interpair variances for MZ and DZ pairs differed, which, of course, may have influenced the correlations, although the figures were not provided for comparison. This innovative study provided interesting material for replication in a larger sample, with additional measures, because of its unexpected assertion that obsessive personality, as here measured, seemed unrelated to hereditary factors. Previous findings of negative correlations between the Lazare scale obsessionality factor and a standard neuroticism measure indicate that the Lazare scale may be measuring obsessionality in a slightly different way from other such questionnaires.

Clifford et al. (1984) used a different measure of obsessionality, the 42-item version of the Leyton Obsessional Inventory (LOI) in conjunction with the Eysenck Personality Questionnaire (EPQ), in a study of 419 pairs of normal volunteer twins from the Institute of Psychiatry Register, with satisfactory zygosity diagnosis (Kasriel & Eaves, 1976). These twins showed the usual preponderance of females and MZs found in volunteer twin registries (Lykken, 1978).

By using a maximum likelihood method on the between- and within-pair mean squares derived from the analysis of variance of the twins' scores on the LOI, much more information was extracted from the data than if simple correlations had been used. The strength of such a quantitative genetic approach is that it is developed from the paradigm of Mendelian genetics and allows hypotheses about genetic architecture to be tested.

Clifford et al. observed that the heritability estimates for trait and symptom scales of the LOI were so similar (47% and 44%, respectively) that they might both be measuring an underlying neurosis variable, so they examined the relationship of the scores on the LOI to the same twins' responses to the EPQ. Examination of the factor structure of these suggested two genetic factors, one influencing a "neuroticism" trait and accounting for a large part of the correlation between EPQ scores and LOI symptoms scores, and an independent factor relating to obsessional traits like "incompleteness," "checking," and to an extent, "cleanliness." Although the influence of heredity on twin resemblance was greater than that of the environment, the latter showed an absence of general factors and in particular of any noticeable effect from the common home environment.

Conclusions

The study of obsessionality, from a disorder or personality viewpoint, is beset by problems. The difficulty of defining the complaints in the disorder remains as salient as ever, and the ensuing wide variation in population prevalence rates has severe implications for studies attempting to quantify the influence of heredity. Until recently the measurement of symptoms and traits was unreliable, but there are now a variety of questionnaire measures available (Cooper & Kelleher, 1973; Goodman et al., 1989; Rachman & Hodgson, 1980; Sanavio, 1988) for comparison with standard neuroticism measures. As yet, geneticists have not fully exploited these instruments.

We concur with some earlier reviewers that it is too soon to conclude that there is a substantial genetic contribution to obsessionality, particularly OCD, though the studies are suggestive of, perhaps, a liability to neurosis, the specific symptoms being due to individual experience. The study of Clifford et al. (1984) does suggest the combined influences of genetic effects on the neuroticism dimension and a smaller specific obsessionality effect, but needs replication. It would also be interesting to see a genetic analysis of other measures, both physiological and behavioral, explored in relation to obsessionality. Finally, any future studies must include an assessment of the role of environmental factors and how they interact with transmissible factors.

References

Alanen, Y. O., Rekola, J. K., Stewen, A., Takala, K. & Tuovinen, M. (1966) The family in the pathogenesis of schizophrenic and neurotic disorders. *Acta Psychiatrica Scandinavica* 42 (Suppl. 189).

Bartlett, D. C. & Steele, J. B. (1979) *Empire: The Life, Legend and Madness of Howard Hughes.* London: Andre Deutsch.

Bille, M. & Juel-Nielsen, N. (1968) Incidence of neurosis in psychiatric and other medical services in a Danish county. *Danish Medical Bulletin* 10:172–176.

Black, A. (1974) The natural history of obsessional states. In *Obsessional States*, ed. H. R. Beech, pp. 20–54. London: Methuen.

Braconi, L. (1970) La psicnevrosi ossessiva nei gemelli. *Acta Geneticae Medicae et Gemellologiae* 19:318–321.

Brown, F. W. (1942) Heredity in the psychoneuroses. *Proceedings of the Royal Society of Medicine* 35:785–790.

Carey, G. (1978) A Clinical Genetic Twin Study of Obsessive and Phobic States. Ph.D. Thesis, University of Minnesota.

Carey, G. & Gottesman, I. I. (1981) Twin and family studies of anxiety, phobic, and obsessive disorders. In *Anxiety: New Research and Changing Concepts*, ed. D. F. Klein & J. Rabkin. New York: Raven Press.

Carey, G., Gottesman, I. I. & Robins, E. (1980) Prevalence rates for the neuroses: Pitfalls in the evaluation of familiality. *Psychological Medicine* 10:437–443.

Clifford, C. A., Murray, R. M. & Fulker, D. W. (1984) Genetic and environmental influences on obsessional traits and symptoms. *Psychological Medicine* 14:791–800.

Comings, D. E. & Comings, B. G. (1987) Hereditary agoraphobia and obsessive-compulsive behaviour in relatives of patients with Gilles De la Tourette's Syndrome. *British Journal of Psychiatry* 151:195–199.

Cooper, J. E. & Kelleher, M. J. (1973). The Leyton Obsessional Inventory: A principle components analysis on normal subjects. *Psychological Medicine* 3:204–208.

Coryell, W. (1981) Obsessive compulsive disorder and primary unipolar depression: Comparisons of background, family history, course and mortality. *Journal of Nervous and Mental Disease* 169:220–224.

Cowie, V. (1961) The incidence of neurosis in the children of psychotics. *Acta Psychiatrica Scandinavica* 37:37–71.

Eaves, L. J., Eysenck, H. J. & Martin, N. G. (1989) *Genes, Culture & Personality: An Empirical Approach.* London: Academic Press.

Emmelkamp, P. M. G. (1983) *Phobic and Obsessive-Compulsive Disorders: Theory, Research and Practice.* New York: Plenum Press.

Eysenck, H. J. & Eysenck, S. B. G. (1964) *Manual of the Eysenck Personality Inventory.* London: University of London.

Goodman, W. K., Price, L. H., Rasmussen, S. A., Mazure, S. A.,

Fleischmann, R. L., Hill, C. L., et al. (1989) The Yale-Brown Obsessive Compulsive Scale. I. Development, use and reliability. *Archives of General Psychiatry* 46:1006–1011.

Gottesman, I. I. & Shields, J. S. (1972) *Schizophrenia and Genetics: A Twin Study Vantage Point.* New York: Academic Press.

Hare, E., Price, J. & Slater, E. (1972) Fertility in obsessional neurosis. *British Journal of Psychiatry* 121:197–205.

Hoaken, P. C. S. & Schnurr, R. (1980) Genetic factors in obsessive-compulsive neurosis? A rare case of discordant monozygotic twins. *Canadian Journal of Psychiatry* 25:167–172.

Ihda, S. (1960) A study of neurosis by the twin method. *Psychiatria et Neurologia Japonica* 63:861–892.

Inouye, E. (1965) Similar and dissimilar manifestations of obsessive-compulsive neurosis in monozygotic twins. *American Journal of Psychiatry* 121:1171–1175.

Insel, T. R., Hoover, C. & Murphy, D. L. (1983) Parents of patients with obsessive compulsive disorder. *Psychological Medicine* 13:807–811.

Jenike, M. A. (1989) Obsessive-compulsive and related disorders: A hidden epidemic. *New England Journal of Medicine* 321:539–541.

Karno, M., Golding, J. M., Sorenson, S. B. & Burman, M. A. (1988) The epidemiology of obsessive-compulsive disorder in five U.S. communities. *Archives of General Psychiatry* 45:1094–1099.

Kasriel, J. & Eaves, L. J. (1976) The zygosity of twins: Further evidence on the agreement between diagnosis by blood groups and written questionnaires. *Journal of Biosocial Science* 8:263–266.

Kendler, K. S., Heath, A. C., Martin, N. G. & Eaves, L. J. (1987) Symptoms of anxiety and symptoms of depression: Same genes, different environments? *Archives of General Psychiatry* 44:451–457.

Kim, S. W., Dysken, M. W. & Kline, M. D. (1990) Monozygotic twins with obsessive-compulsive disorder. *British Journal of Psychiatry* 156:435–438.

Kringlen, E. (1965) Obsessional neurotics: A long term follow-up. *British Journal of Psychiatry* 111:709–722.

Lazare, A., Klerman, G. L. & Armor, D. J. (1970) Oral, obsessive and hysterical personality patterns: Replication of factor analysis in an independent sample. *Journal of Psychiatric Research* 7:275–290.

Lemkau, P., Tietze, C. & Cooper, M. (1942) Mental hygiene problems in an urban district. *Mental Hygiene* 26:100–119.

Lewis, A. J. (1936) Problems of obsessional illness. *Proceedings of the Royal Society of Medicine* 29:325–336.

Lewis, A. J. (1938) The diagnosis and treatment of obsessional states. *Practitioner* 141:21–30.

Lewis, A. J. (1957) Obsessional illness. *Acta Neuropsiquiatrica Argentina* 3:323–335.

Lewis, S., Chitkara, B. & Reveley, A. M. (1991, in press) Obsessive-compulsive disorder and schizophrenia in three identical twin pairs. *Psychological Medicine* 21(1).

Lo, W. H. (1967) A follow-up study of obsessional neurotics in Hong Kong Chinese. *British Journal of Psychiatry* 113:823–832.

Luxenberger, H. (1930). Hereditat und Familientypus der Zwangsneurotiker. *Archiv fur Psychiatrie* 91:590–594.

Lykken, D. T. (1978) Volunteer bias in twin research: The rule of two thirds. *Social Biology* 25:1–9.

Macdonald, A. M. & Murray, R. M. (1989) A twin study of obsessive compulsive neurosis. Presented at the 6th International Congress on Twin Studies, 28–31 August, Rome.

Mahgoub, O. M., Ahmed, M. A. M. & Al-Suhaibani, M. O. (1988) Identical Saudi twins concordant for obsessive-compulsive disorder. *Saudi Medical Journal* 9:641–643.

Mapother, E. & Lewis, A. (1941) Obsessional disorder. In *A Textbook of the Practice of Medicine*, 6th ed., ed. F. W. Price. London: Oxford University Press.

Marks, I. M., Crowe, M., Drewe, E., Young, J. & Dewhurst, W. G. (1969) *British Journal of Psychiatry* 115:991–998.

McGuffin, P. & Mawson, D. (1980) Obsessive-compulsive neurosis: Two identical twin pairs. *British Journal of Psychiatry* 137:285–287.

McKeon, P. & Murray, R. M. (1987) Familial aspects of obsessive-compulsive neurosis. *British Journal of Psychiatry* 151:528–534.

Parker, N. (1964a) Twins: A psychiatric study of a neurotic group. *Medical Journal of Australia* 2:735–742.

Parker, N. (1964b) Close identification in twins discordant for obsessional neurosis. *British Journal of Psychiatry* 110:496.

Pauls, D. L. (1989) The familial relationship of obsessive compulsive disorder and Tourette's syndrome. Summary in *Proceedings of the 142nd APA Annual Meeting*.

Rachman, S. J. & De Silva, P. (1978) Abnormal and normal obsessions. *Behaviour Research and Therapy* 16:233–248.

Rachman, S. J. & Hodgson, R. J. (1980) *Obsessions and Compulsions*. Englewood Cliffs, NJ: Prentice-Hall.

Rapoport, J., Elkins, R., Langer, D. H., Sceery, W., Buchsbaum, M. S., Gillin, J. C., et al. (1981) Childhood obsessive-compulsive disorder. *American Journal of Psychiatry* 138:1545–1555.

Rasmussen, S. A. & Tsuang, M. T. (1986) Clinical characteristics and family history in DSM-III obsessive-compulsive disorder. *American Journal of Psychiatry* 143:317–322.

Reed, G. F. (1985) *Obsessional Experience and Compulsive Behaviour: Cognitive-Structural Approach*. London: Academic Press.

Reich, T., James, J. W. & Morris, C. A. (1972) The use of multiple thresholds in determining the model of transmission of semi-continuous traits. *Annals of Human Genetics* 36:163–184.

Robins, L. N., Helzer, J. E., Croughan, J. & Ratcliff, K. S. (1981) National Institute of Mental Health Diagnostic Interview Schedule: Its history, characteristics and validity. *Archives of General Psychiatry* 38:381–389.

Rosenberg, C. M. (1967) Familial aspects of neurosis. *British Journal of Psychiatry* 113:405–413.

Rudin, E. (1953) Ein Beitrag zur Frage der Zwangskrankheit, insbesondere ihrer hereditaren Beziehungen. *Archiv fur Psychiatrie und Zeitschrift Neurologie* 191:14–54.

Rutter, M. (1966) *Children of Sick Parents*. London: Oxford University Press.

Sanavio, E. (1988) Obsessions and compulsions: The Padua Inventory. *Behaviour Research and Therapy* 26:169–177.

Schepank, H. (1976) Heredity and environmental factors in the development of psychogenic diseases. *Acta Geneticae Medicae et Gemellologiae* 25:237–239.

Scott, A. M., Kelleher, M. J., Smith, A. & Murray, R. M. (1982) Regional differences in obsessionality and obsessional neurosis. *Psychological Medicine* 12:131–134.

Shapiro, A. K. & Shapiro, E. (1989) Association of obsessive compulsive disorder with Tourette's syndrome: Fact or artifact? Summary in *Proceedings of the APA 142nd Annual Meeting*.

Smith, C. A. B. (1970) Heritability of liability and concordance in monozygotic twins. *Annals of Human Genetics* 34:85.

Tarsh, M. J. (1978) Severe obsessional illness in dizygotic twins treated by leucotomy. *Comprehensive Psychiatry* 19:165–169.

Tienari, P. (1963) Psychiatric illnesses in identical twins. *Acta Psychiatrica Scandinavica* 39(Suppl. 171).

Torgersen, S. (1980) The oral, obsessive, and hysterical personality-syndromes. A study of hereditary and environmental factors by means of the twin method. *Archives of General Psychiatry* 37:1272–1277.

Torgersen, S. (1983) Genetic factors in anxiety disorders. *Archives of General Psychiatry* 40:1085–1089.

Tyrer, P. (1985) Neurosis divisible? *Lancet* i:685–688.

Vaisanen, E. (1976) Psychiatric disorders in Finland. In *Social Somatic and Psychiatric Studies of Geographically Defined Populations*, ed. T. Anderson, C. Astrup & A. Forsdahl, pp. 22–23. *Acta Psychiatrica Scandinavica* 263 (Suppl.).

Walton, H. J. & Presly, A. S. (1973) Use of a category system in the diagnosis of abnormal personality. *British Journal of Psychiatry* 122:259–268.

Woodruff, R. & Pitts, F. N. (1964) Monozygotic twins with obsessional illness. *American Journal of Psychiatry* 120:1075–1080.

Young, J. P. R., Fenton, G. W. & Lader, M. H. (1971) The inheritance of neurotic traits. *British Journal of Psychiatry* 119:393–398.

Part 3

Examples and Applications

Chapter 11
Genetic Latent Structure Models

STEVEN MATTHYSSE

This chapter summarizes explorations with genetic latent structure models that have been reported elsewhere, but not so far gathered in one place. Details may be found in the references.

The Concept of Genetic Latent Structure Models

This class of models arises by taking as fundamental the idea of *pleiotropy*, the tendency of genes to have multiple phenotypic effects. Variations in structural proteins, enzymes, and receptors will obviously affect many traits, since these substances are general machinery of the nervous system. Other known mechanisms of gene action in the brain are also pleiotropic; for example, regulation of neuron number (Williams & Herrup, 1988) and control of cell migration during embryogenesis (Caviness & Rakic, 1978). The numbers of dopamine-containing neurons in the substantia nigra and in the ventral tegmental area are under coordinated genetic control (Reis et al., 1983). Because the cell numbers in the two regions are correlated with each other, variations in these loci will simultaneously affect both motor and emotional behavior. In the Reeler mouse mutant, it has been shown that outward migration of neurons from germinal layers is under the influence of a single gene that determines whether the migrating neurons detach at the right place from the glial fibers that serve as conduits (Pinto-Lord et al., 1982). Both the cerebral and cerebellar cortices are affected by the mutation.

To be sure, gene effects in the central nervous system of entirely different types may be discovered as research in this field progresses; but so far the generalization appears valid that single genes affect multiple brain systems and behavioral processes. Therefore we have taken pleiotropy as a starting-point for a genetic model applicable to mental disorders.

Genetic latent structure models have the following general form:

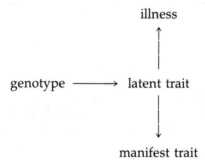

The latent trait is considered unobservable only relative to the present state of our knowledge; it refers to an unknown but real process that causes both the illness and the associated manifest trait.

Formal Specification of the Model

In principle, there can be continuous or discrete states of the latent trait, manifest trait, and illness; if the variables are categorical, the number of categories can be large or small; the genotype can be a single locus, several loci, or polygenic. In addition, the model can become multilayered, the manifest trait in the diagram being regarded as a latent trait with paths to several other observed traits, and so on. Naturally, there is a danger in making these models too complicated, because their parameters will be subject to large standard errors, and may even become unidentifiable. We have found the simplest possible model of this type, in which the latent trait, the manifest trait, and the illness are regarded as either present or absent, to be useful in empirical work.

In our studies the genotype has been taken to be a single locus. We regarded models with multiple discrete loci as likely to be unidentifiable for practical purposes, given the data presently available in psychiatric genetics. A polygenic interpretation of the genotype would be simple enough to be workable, but we avoided it because we wanted our model to serve as an adjunct to linkage analysis.

An objection naturally arises to decisions of this kind. If a single locus model is chosen in preference to a polygenic model because the model is intended for use in linkage analysis, rather than because segregation or recurrence risk data favor the single locus, the assumptions of the linkage analysis seem to prejudge the result. On the other hand, it is agreed by most observers that decisions between discrete and continuous loci cannot be effectively made without the aid of data from linkage experiments. Segregation and recurrence risk data are not powerful enough to make the discrimination between realistic forms of these models. Paradoxically, we seem to need a major locus model to carry out linkage analysis, but we cannot validate that model before we do the analysis. We shall return to this issue later.

If the latent trait gene is assumed to have two alleles, there are three genotypes, in which the pathogenic allele is absent, heterozygous, or homozygous. The probability that an individual has the latent trait is assumed to depend on his genotype, π_1, π_2, and π_3 being the probabilities if the pathogenic allele is absent, heterozygous, or homozygous, respectively. The probability of observing the illness depends on the presence of the latent trait: r_I^+ if it is present, r_I^- if it is absent. Similarly, the probability of observing the manifest trait is r_T^+ if the latent trait is present, r_T^- if it is absent. The empirical data predicted by the model can be arranged in a $3 \times 2 \times 2$ table, $\phi_g(m,i)$ representing the frequency of observing each manifest trait-illness combination for each genotype. The index g refers to the genotype, m to the manifest trait, and i to the illness; for m and i, 1 means absent, 2 present. The $\phi_g(m,i)$ are not directly observed, but are inferred from transmission patterns within pedigrees. Unless otherwise specified, when we speak of the latent model (LT), we shall mean this specific formulation.

The equations of LT are:

$$(1 - \pi_g)(1 - r_T^-)(1 - r_I^-) + \pi_g(1 - r_T^+)(1 - r_I^+) = \phi_g(1,1)$$
$$(1 - \pi_g)\, r_T^-\, (1 - r_I^-) + \pi_g r_T^+(1 - r_I^+) = \phi_g(2,1) \qquad \text{(LT)}$$
$$(1 - \pi_g)(1 - r_T^-)r_I^- + \pi_g(1 - r_T^+)r_I^+ = \phi_g(1,2)$$

for $1 \le g \le 3$. The first equation, for example, expresses the fact that the probability that an individual with genotype g has neither the manifest trait nor the illness is the sum of two terms, each corresponding to a possible latent trait state. The first term corresponds to the possibility that he does not have the latent trait; the second term, to the possibility that he does. The first term is the product of three factors: $(1 - \pi_g)$, the probability the individual does not have the latent trait; $(1 - r_T^-)$, the probability that he does not have the manifest trait, conditioned on his not having the latent trait; and $(1 - r_I^-)$, the probability he does not have the illness, conditioned on his not having the latent trait. The second term is constructed in a similar way, as are the other equations. An equation for $\phi_g(2,2)$ is not necessary because the corresponding quantities all add to 1. To avoid ambiguity, we adopt the convention $r_T^+ \geq r_T^-$, because if $r_T^+ < r_T^-$, the transformation $\pi_g \rightarrow 1 - \pi_g$ ($1 \leq g \leq 3$), $r_T^+ \leftrightarrow r_T^-$, $r_I^+ \leftrightarrow r_I^-$ leads to a model that makes exactly the same predictions and satisfies $r_T^+ > r_T^-$; the transformation amounts to renumbering the states of the latent trait.

LT Is Falsifiable

Whenever a model has a large number of adjustable parameters, the question arises whether it is falsifiable; in this context, whether any data consistent with major locus transmission of the manifest trait and the illness would be inconsistent with the model. Otherwise the model might be a useful way of summarizing and interpreting observations, but pedigree data could never be said to confirm or disconfirm it. By manipulating the equations of LT, one can show that certain ratios must hold between the $\phi_g(l,m)$ in order that the system (LT) have any solution at all:

$$\frac{\phi_1(1,1) - \phi_2(1,1)}{\phi_1(1,1) - \phi_3(1,1)} = \frac{\phi_1(2,1) - \phi_2(2,1)}{\phi_1(2,1) - \phi_3(2,1)} = \frac{\phi_1(1,2) - \phi_2(1,2)}{\phi_1(1,2) - \phi_3(1,2)}$$

Since these ratios may not be observed, LT is falsifiable. In general, one can show that genetic latent structure models with G genotypes, L states of the latent trait, M states of the manifest trait, and I states of the illness are falsifiable if $GMI > L(G + M + I - 2)$, although (as mentioned above) statistical problems may arise if G, L, M, or I are large. For an extended discussion of falsifiability, see Matthysse (in press).

Relationships With Other Models

The closest relation to genetic latent structure models is the family of latent structure models proposed by Goodman and others (Goodman, 1974). The hallmark of these models is that the correlation between the observed variables disappears when the status of the latent trait is held constant. They cannot be tested directly through prediction, however, if the latent trait is not observed. Genetic latent structure models are essentially latent structure models in Goodman's sense, with a genetic dimension added to the latent trait.

If we take away the genetic aspect, the latent structure model becomes:

$$\text{manifest trait} \leftarrow \text{latent trait} \rightarrow \text{illness}$$

Setting $G = 1$ (since effectively everyone has the same genotype if genetic effects are not considered) the nongenetic latent structure model is falsifiable if $MI > L(M + I - 1)$, where M is the number of states of the manifest trait, I is the number of states of the illness, and L is the number of states of the latent trait. In the simplest case $L = M = I = 2$, this inequality reduces to $4 > 6$, so the system is underdetermined. The case $L = 2$, $M = I = 3$ is also underdetermined; but versions with more discriminable states of the manifest trait and illness do have more equations than unknowns. For example, the inequality holds if $L = 2$, $M = I = 4$. When data are subject to interfering factors that cannot be directly measured, however, versions of the latent trait model with more states of the manifest trait and illness may not effectively solve the problem of underdetermination, because the more states are allowed, the more likely misclassification problems are to occur for the manifest trait, and the more likely accidental factors are to move the patient unpredictably from one illness state to another. An adequate model would have to include parameters for these processes, so it might still end up with excess degrees of freedom.

Ideas similar to the latent trait concept also underlie the "taxometric" proposals of Meehl and Golden (Meehl, 1973). The taxometric program is oriented toward the attempt to detect major gene effects by purely statistical means (Golden, in press), whereas we have envisaged the latent trait model as an adjunct to linkage analysis.

Diathesis and Epiphenomenon Submodels

When $L = M$ or $L = I$, the latent structure model contains interesting submodels as special cases. They are the *diathesis* and *epiphenomenon* models. A diathesis is "a permanent (hereditary or acquired) condition of the body which renders it liable to certain special diseases or affections; a constitutional predisposition or tendency" (Oxford English Dictionary, 1971). An epiphenomenon is "something that appears in addition; a secondary symptom" (ibid.). To construct the diathesis submodel as a special case of the latent trait model, let $r_T^- = 0, r_T^+ = 1$; then the manifest trait becomes synonymous with the latent trait, and the causal diagram collapses into the diathesis model. Similarly, to construct the epiphenomenon model, let $r_I^- = 0, r_I^+ = 1$; then the illness becomes synonymous with the latent trait, and the diagram collapses into the epiphenomenon model.

$$\text{genotype} \rightarrow \text{manifest} \rightarrow \text{illness (diathesis)}$$
$$\text{trait}$$

$$\text{genotype} \rightarrow \text{illness} \rightarrow \text{manifest}$$
$$\text{trait (epiphenomenon)}$$

The diathesis and epiphenomenon submodels are falsifiable with respect to each other, as the following argument shows. If $r_T^- = 0$ and $r_T^+ = 1$ (the defining conditions for the diathesis submodel), equations (LT) become:

$$(1 - \pi_g)(1 - r_I^-) = \phi_g(1,1) \qquad \pi_g(1 - r_I^+) = \phi_g(2,1)$$
$$(1 - \pi_g)r_I^- = \phi_g(1,2)$$

so that $\dfrac{\phi_g(1,1)}{\phi_g(1,2)}$ is independent of g. All risks in relatives are determined by the subject's genotype, so the constancy of this ratio implies that risks in relatives for either the manifest trait or the illness will not depend on the proband's illness status, if his trait status is held fixed. This is a strong prediction, which will not be true in general of the epiphenomenon model, as can be seen by examining the corresponding equations, setting $r_I^- = 0$ and $r_I^+ = 1$ in (LT):

$$(1 - \pi_g)(1 - r_T^-) = \phi_g(1,1) \quad (1 - \pi_g)r_T^- = \phi_g(2,1)$$
$$\pi_g(1 - r_T^+) = \phi_g(1,2)$$

The argument works the same way in the other direction.

Empirical Studies

The major empirical application of the genetic latent trait model to date has been to smooth pursuit visual tracking and schizophrenia. The smooth pursuit eye movement (SPEM) system is dysfunctional in a large proportion of schizophrenic patients. Normally, when the pursuit system is turned on, as it is when a person is following a moving target, the saccadic, or rapid eye movement, system is turned off. In about 65% of schizophrenic patients and about 40% of their first-degree relatives (but only in about 8% of normal people) saccadic events intrude into the pursuit movements. Some of the time SPEM is replaced completely by saccadic eye movements; at other times, SPEM is interrupted by saccadic events, some of which attempt to correct pursuit that is too slow to keep up with the velocity of the target, and some of which are simply saccadic intrusions. The saccadic system itself generally functions well in schizophrenic patients. The smooth pursuit disturbance in schizophrenia is not an artifact of inattention or antipsychotic drug effects (see Holzman, 1985, for a review).

Genetic modeling of eye movement disturbances (EMDs) and schizophrenia was stimulated by the finding that these disturbances also occur in about 40% of the first-degree relatives of schizophrenic patients (even when the relatives have no history of major mental disorder themselves), but only in 14% of the first-degree relatives of patients with affective disorder. Abnormal pursuit in a schizophrenic proband tends to be associated with abnormal pursuit in at least one of the two clinically unaffected parents (Holzman & Levy, 1977). In MZ and DZ twins who are discordant for schizophrenia, concordance for pursuit dysfunctions was over 80% in the MZ sets and over 40% in the DZ sets (Holzman et al., 1977; Holzman et al., 1980).

The LT model was fitted to data on transmission patterns of schizophrenia and EMDs in a sample of nuclear families obtained in Chicago and Boston (Matthysse et al., 1986). Later it was shown that the same parameters were consistent with data obtained in an independent study using a Norwegian sample of offspring of MZ and DZ twins discordant for schizophrenia (Holzman et al., 1988).

The latent trait allele, according to the maximum-likelihood parameters, is a common dominant (population frequency 3.8%).

The penetrance of this allele for clinical schizophrenia, although ten times the population prevalence, is only 7.1%. In other words, the model is not strictly a single major locus model for schizophrenia, but is more accurately interpreted as multigenic; the postulated locus exerts only a modest degree of control over the clinical phenotype. The remainder of the determination might be from other loci, environment, or chance. From the point of view of schizophrenia alone, the postulated gene is a minor locus. When schizophrenia and EMDs are considered together as two independent manifestations of a latent trait, however, the data are consistent with the latent trait being genetically transmitted as an autosomal dominant with high penetrance.

Both the diathesis and epiphenomenon submodels can be ruled out as explanations of the relationship between EMDs and schizophrenia. The diathesis model, as explained above, predicts that the frequency of bad tracking, in the relatives of schizophrenic patients whose tracking is normal, should be about the same as in the general population. Empirically, the opposite is found; schizophrenic patients with normal pursuit frequently have healthy relatives with impaired pursuit. Indeed, in several sets of clinically discordant DZ twins, the schizophrenic twin had normal pursuit and the healthy twin had impaired pursuit. The epiphenomenon model gives a significant χ^2 for deviation from goodness of fit in schizophrenia, in comparison to the LT model, whereas the relationship between EMDs and affective disorder can be accounted for as an epiphenomenon (Matthysse et al., 1986). The LT model has also been applied to attention deficit disorder (Deutsch et al., 1990).

Epistemological Status of the Model

Fitting the parameters of a model by maximum-likelihood methods is not a test of the validity of the model. The replication of the maximum-likelihood parameters in another sample is valuable because, if the parameters were inconsistent across samples, it would be difficult to make practical use of the model. Nevertheless replication does not, strictly speaking, contribute to verification, because a false model might also give consistent parameter fits if the actual transmission process were constant across samples.

In the Chicago-Boston study, the LT model was incorporated in a more inclusive model that allowed the probability that hetero-zygotes transmit the pathogenic allele to their offspring to be a free parameter, rather than being fixed at its Mendelian value of 0.5. The maximum-likelihood estimates of the segregation ratio was 0.52, the deviation from 0.5 being nonsignificant. This test con-tributes some confirmatory evidence, but it is not very stringent. We have not carried out extensive tests of the model, because we believe convincing tests are impossible without linkage data.

The recurrence risks to monozygotic twins and offspring of dual matings are underestimated by the latent trait model for schizophrenia and EMDs, with the parameters derived from the Chicago-Boston and Norway samples. McGue and Gottesman (1989) have argued that this limitation invalidates the model. We consider the discrepancy to be of no fundamental importance, be-cause it could easily be eliminated by incorporating additional genetic or shared environmental effects influencing the develop-ment of schizophrenia from the latent trait (Risch, 1990a). These effects might have more to do with general adaptation than with the specific etiology of schizophrenia. It would be important to include them in a model designed for application to a sample in-cluding dual matings or identical twins, but these effects can probably be safely ignored when the model is applied to test for linkage in pedigrees without twins or dual matings.

Incompletely Validated Models in Tests of Linkage

The debate over monozygotic twin and dual-mating recurrence risks exemplifies the weakness of recurrence risks and segrega-tion data to validate or invalidate any but the simplest models of the genetics of mental illness. Transmission data may be inconsis-tent with a single Mendelian gene, but melt into consistency if a second modifying gene is postulated; a linear model may rule out family environment effects, but add a nonlinear term, and they become plausible again.

This pessimistic view of the power of family transmission data to differentiate between genetic models is counterbalanced by a more optimistic view of the potential of linkage analysis. This view, also, is widely shared. If a mental illness turns out to be linked to a gene on a specific chromosome, we know decisively that at least one major gene is involved, and there is hope that the gene-product can be isolated and a systematic investigation of the causal chain begun.

Given the current state of knowledge in psychiatric genetics, it would be understandable to have a low opinion of mathematical models and a high opinion of linkage analyses. The difficulty with this position is that, if linkage analysis is restricted to model-free methods, such as the Penrose sib-pair technique (Penrose, 1935) and its modern extensions (Weeks & Lange, 1988), it is not likely to be powerful enough to map the low-penetrance loci that we expect in mental diseases like schizophrenia (Risch, 1990b). We shall see this later in the case of the latent trait model applied to schizophrenia. Models cannot be validated without the data provided by linkage analyses, but linkage analyses cannot be carried out effectively without models.

This dilemma can be escaped if we regard models, not as postulates about nature which have to be judged true or false, but as generators of statistical tests to be used in linkage analysis. From this point of view, models are detectors of deviation from an independent assortment of marker genotypes and trait phenotypes. The first null hypothesis in linkage analysis is that market genotypes are distributed independently of trait phenotypes. If the null hypothesis cannot be rejected for a particular marker, there is no point in pursuing that marker further. Although rejection of the null hypothesis does not prove that the marker and the trait are linked (there could, for example, be association effects), it justifiably highlights that marker for further studies.

There are many ways of constructing a statistical test of the hypothesis that marker genotypes are distributed independently of trait phenotypes. Indeed, any mathematical model of the transmission process that includes a discrete locus can be used to define such a test. The method is familiar: using the model, compute the likelihood of the joint marker-trait distribution in the sample of pedigrees, on the assumption that the recombination fraction is .5; compute the maximum likelihood as the recombination fraction is allowed to vary freely between .5 and 1; take the logarithm of the odds ratio. Accept the null hypothesis if the lod score is below a specified threshold, reject the null hypothesis if it is above. Our concern becomes focused on the statistical properties of the test the model generates, rather than on the truth or falsity of the assumptions of the model. A false model can generate a valid test of deviation from independence, providing the threshold is properly set. The price we pay for using a false model is that it will usually be less powerful a test than a true model.

This idea seems contrary to usual ways of thinking, so we will give an example. Suppose we want to test the hypothesis that birds' nests in our woods are distributed nonrandomly. We can lay out 6-foot wide strips and count the number of nests in each strip, then test for deviation from a Poisson distribution. Such a test would be a poor choice, because it is not likely that this model reflects the actual deviation from randomness in the pattern of birds' nests. A model associating nests with certain species of trees would be more likely to be true, and a statistical test of nonrandomness based on the tree-species model would have greater power. Nevertheless there is nothing logically wrong with the 6-foot-strip test, and it might work. All that is required for it to be a valid test is that we choose a criterion level that correctly reflects the probability of false positives if the birds' nest distribution is actually random. Agreement with nature is necessary for a model to have power to reject the null hypothesis, but correct setting of the threshold is all that is necessary for it to generate a valid statistical test.

Computing the probability of false positives in the linkage context is straightforward. Assume that we are working with a sample of pedigrees in which the trait phenotypes of each individual have already been determined. Generate sets of marker genotypes by Monte Carlo simulation according to Mendelian rules, ignoring the trait phenotypes, and plot a histogram of lod scores obtained when the null hypothesis is true. Set a threshold that gives an acceptable rate of false positives.[1]

Estimating the probability of false negatives is more subtle, because the concept of a "false negative" only acquired meaning in the context of some alternative hypothesis. For example, we can ask: "if the actual state of affairs were exactly the one postulated by the model, with a recombination fraction θ ($\neq$ .5), what is the probability that the lod score would exceed threshold?" That is the usual question in power analysis. On the other hand, we could just as well ask how often the threshold would be exceeded if some other model were true, not the one adopted for the statistical test. In principle, it might turn out that a more powerful test could be constructed by deliberately setting the parameters of the model to different values from the maximum-likelihood estimates arrived at in previous studies, even if those estimates were used to define the alternative hypothesis. If that happened, the more powerful model, not the maximum-likelihood model, should be used in the first steps of linkage analysis.

When a model is used in this way, we have to resist the temptation of interpreting success in detecting linkage as confirmation of the model. Detection of linkage offers a little bit of evidence in favor of the model, since models that agree with reality have more power than those that do not, but it would be risky to depend much on this evidence. The purpose of the model, at this stage of the game, is only to provide a tool for detecting nonindependence of marker genotypes from trait phenotypes. That will be enough to push open the Gate of Molecular Biology, which is what we really wanted our model for in the first place.

The Power of Linkage Analysis in Schizophrenia, According to the LT Model

Perhaps the most significant application of genetic latent structure models is a set of simulation studies, as yet largely unpublished, suggesting that if the latent trait model for schizophrenia is true, linkage analysis that takes schizophrenia alone into account as the phenotype is virtually guaranteed to fail.

The need to assess the power of linkage analysis in schizophrenia arose from a practical situation: there are in Denmark several large schizophrenia pedigrees available for linkage studies, but the number of clinically affected individuals in each pedigree is small, as it is in most families. In order to estimate the power of linkage analysis, we assumed the LT model as the alternative hypothesis, with the parameters derived from our Norwegian study of offspring of discordant twins (Holzman et al., 1988). The recombination fraction between the marker and the latent trait locus was assumed to be .05.

We used a recently developed random walk method for simulating the distribution of trait and marker genotypes in pedigrees where trait phenotypes are already known (Lange & Matthysse, 1989).[2] Although it is easy to generate trait phenotype and marker genotype vectors randomly for a pedigree, it is not practical to select from those simulations the ones that reproduce the pattern of observed trait phenotypes. The number of combinations that would have to be rejected is many orders of magnitude too large. The method we used is an adaptation of the Metropolis algorithm in statistical physics. The crux of the method is the construction of a random walk with equilibrium distribution matching the conditional probabilities of the latent trait and marker genotype vector, given the observed phenotypes. Each sample from the random walk defines a set of latent trait and marker genotype vectors conditioned on the observed schizophrenia phenotypes. Since the eye tracking phenotype had not yet been observed, we conditioned only on the presence or absence of schizophrenia; eye tracking phenotypes were generated in accordance with the latent trait model by an additional simulation step.

The marker genotypes, simulated eye tracking phenotypes, and observed schizophrenia phenotypes were used as input for linkage analysis. In the linkage analysis step, calculations can be based either on schizophrenia alone as the observed phenotype, or on both schizophrenia and eye tracking, taking both into account by means of the latent trait model. Our calculations show that, if LT is true, the expected lod scores when observations are restricted to schizophrenia alone are very low, whereas a considerably more optimistic picture emerges when both phenotypes are taken into account. This outcome is understandable, since restriction of the observations to schizophrenia alone provides many fewer informative individuals. If the latent trait model is true, taking into account both phenotypes could make the difference between success and failure in linkage analysis of schizophrenia.

Extensions of the Latent Trait to Other Phenotypes

It seems implausible that a latent trait related to a major gene would have just schizophrenia and EMDs as effects, and no others. Except for schizophrenia spectrum disorders (Kendler et al., 1981; Kety et al., 1978), it is surprising how little has been done to investigate psychological processes other than eye tracking in family members of probands with mental illness. A few other phenomena have been studied in this way:

1. The Continuous Performance Task of sustained attention (Erlenmeyer-Kimling & Cornblatt, 1978; Nuechterlein, 1983)
2. Span of apprehension (Wagener et al., 1986)
3. Late components of evoked responses in the EEG (Friedman et al., 1982; Saitoh et al., 1984)

For a review of possible candidates for extension of the latent trait concept, see Holzman and Matthysse (1990). Only in the case of smooth pursuit tracking have quantitative genetic models been constructed. Including other manifest traits in addition to EMDs might increase the power of the latent trait model to detect deviation from randomness in the assortment of marker genotypes and trait phenotypes, and enhance its usefulness in linkage analysis.

Notes

1. The conventionally acceptable rate of false positives for a single marker takes into account the practice of simultaneously testing many markers (Lander & Botstein, 1989) and the fact that the sample size is not fixed in advance (Morton, 1955). The theorem of Wilks (1938) on asymptotic properties of maximum-likelihood estimators provides conditions under which the lod score for a single marker will have a χ^2 distribution under the null hypothesis; but we think it is misapplied in pedigree analysis, because it is asymptotic, and the appropriate sample size is the number of pedigrees, not the number of individuals in a pedigree. Monte Carlo simulation is safer than assuming a χ^2 distribution.

2. Figure 4 in this paper is not correct, because of a computational error that inflated lod scores. After correction the qualitative conclusions remain the same, but the lod scores reached by the LT model are not as high.

References

Caviness, V. S., Jr. & Rakic, P. (1978) Mechanisms of cortical development: A view from mutation in mice. *Annual Review of Neuroscience* 1:297–326.

Deutsch, C. K., Matthysse, S., Swanson, J. M. & Farkas, L. G. (1990) Genetic latent structure analysis of dysmorphology in attention deficit disorder. *Journal of the American Academy of Child and Adolescent Psychiatry* 29:189–194.

Erlenmeyer-Kimling, L. & Cornblatt, B. (1978) Attentional measures in a study of children at high risk for schizophrenia. In *The Nature of Schizophrenia: New Approaches to Research and Treatment*, ed. L. C. Wynne, R. L. Cromwell & S. Matthysse. New York: John Wiley.

Friedman, D., Vaughan, H. G. & Erlenmeyer-Kimling, L. (1982) Cognitive brain potentials in children at risk for schizophrenia: Preliminary findings. *Schizophrenia Bulletin* 8:514–531.

Golden, R. R. (in press) Bootstrapping taxometrics: On the development of a method for detection of a single major gene.

Goodman, L. A. (1974) The analysis of qualitative variables when some of the variables are unobservable. Part I: A modified latent structure approach. *American Journal of Sociology* 79:1179–1259.

Holzman, P. S. (1985) Eye movement dysfunctions and psychosis. *International Review of Neurobiology* 27:179–205.

Holzman, P. S., Kringlen, E., Levy, D. L. & Haberman, S. J. (1980) Deviant eye tracking in twins discordant for psychosis: A replication. *Archives of General Psychiatry* 37:627–631.

Holzman, P. S., Kringlen, E., Levy, D. L., Proctor, L. R., Haberman, S. & Yasillo, N. J. (1977) Abnormal pursuit eye movements in schizophrenia: Evidence for a genetic marker. *Archives of General Psychiatry* 34:802–085.

Holzman, P. S., Kringlen, E., Matthysse, S., Flanagan, S. D., Lipton, R. B., Cramer, G., et al. (1988) A single dominant gene can account for eye tracking dysfunctions and schizophrenia in offspring of discordant twins. *Archives of General Psychiatry* 45:641–647.

Holzman, P. S. & Levy, D. L. (1977) Smooth-pursuit eye movements and functional psychoses: A review. *Schizophrenia Bulletin* 3:15–27.

Holzman, P. S. & Matthysse, S. (1990) Review: The genetics of schizophrenia. *Psychological Science* 1:279–286.

Kendler, K. S., Gruenberg, A. M. & Strauss, J. S. (1981) An independent analysis of the Copenhagen sample of the Danish adoption study of schizophrenia: II. The relationship between schizotypal personality disorder and schizophrenia. *Archives of General Psychiatry* 38:982–984.

Kety, S. S., Wender, P. H. & Rosenthal, D. (1978) Genetic relationships within the schizophrenia spectrum: Evidence from adoption studies. In *Critical Issues in Psychiatric Diagnosis*, ed. R. I. Spitzer & D. F. Klein. New York: Raven Press.

Lander, E. S. & Botstein, D. (1989) Mapping Mendelian factors underlying quantitative traits using RFLP linkage maps. *Genetics* 121:185–199.

Lange, K. & Matthysse, S. (1989) Simulation of pedigree genotypes by random walks. *American Journal of Human Genetics* 45:959–970.

Matthysse, S. (in press) Genetics and the problem of causality in abnormal psychology. In *Comprehensive Handbook of Psychopathology*, 2nd ed., ed. P. B. Sutker & H. E. Adams. New York: Plenum Press.

Matthysse, S., Holzman, P. S. & Lange, K. (1986) The genetic transmission of schizophrenia: Application of Mendelian latent structure analysis to eye tracking dysfunctions in schizophrenia and affective disorders. *Journal of Psychiatric Research* 20:57–67.

McGue, M. & Gottesman, I. I. (1989) Genetic linkage in schizophrenia: Perspectives from genetic epidemiology. *Schizophrenia Bulletin* 15:453–464.

Meehl, P. E. (1973) MAXCOV-HITMAX: A taxonomic search method for loose genetic syndromes. In *Psychodiagnostics: Selected Papers*, ed. P. E. Meehl. Minneapolis: University of Minnesota Press.

Morton, N. E. (1955) Sequential tests for the detection of linkage. *American Journal of Human Genetics* 7:277–315.

Nuechterlein, K. H. (1983) Signal detection in vigilance tasks and behavioral attributes among offspring of schizophrenia mothers and among hyperactive children. *Journal of Abnormal Psychology* 92:4–28.

Oxford English Dictionary (Compact Edition) (1971) Oxford: Oxford University Press.

Penrose, L. S. (1935) The detection of autosomal linkage in data which consist of pairs of brothers and sisters of unspecified parentage. *Annals of Eugenics* 6:133–138.

Pinto-Lord, M. C., Evrard, P. & Caviness, V. S., Jr. (1982) Obstructed neuronal migration along radial glial fibers in the neocortex of the reeler mouse: A Golgi-EM analysis. *Developmental Brain Research* 4:379–393.

Reis, D. J., Fink, J. S. & Baker, H. (1983) Genetic control of the number of dopamine neurons in the brain: Relationships to behavior and responses to psychoactive drugs. *Research Publications of the Association for Research in Nervous and Mental Disease* 60:55–75.

Risch, N. (1990a) Linkage strategies for genetically complex traits. I. Multilocus models. *American Journal of Human Genetics* 46:222–228.

Risch, N. (1990b) Linkage strategies for genetically complex traits. II. The power of affected relative pairs. *American Journal of Human Genetics* 46:229–241.

Saitoh, O., Niwas, S., Hiramatsu, K., Kameyama, T., Rymar, K. & Itoh, K. (1984) Abnormalities in late positive components of event-related potentials may reflect a genetic predisposition to schizophrenia. *Biological Psychiatry* 19:293–303.

Wagener, D. K., Hogarty, G. E., Goldstein, M. J., Asarnow, R. F. & Browne, A. (1986) Information processing and communication disorders in schizophrenia patients and their mothers. *Psychiatry Research* 18:365–377.

Weeks, D. E. & Lange, K. (1988) The affected-pedigree-member method of linkage analysis. *American Journal of Human Genetics* 42:315–326.

Williams, R. W. & Herrup, K. (1988) The control of neuron number. *Annual Review of Neuroscience* 11:423–453.

Wilks, S. S. (1938) The large sample distribution of the likelihood ratio for testing composite hypotheses. *Annals of Mathematical Statistics* 9:60–62.

Chapter 12

Genetic Influences and Criminal Behavior

PATRICIA A. BRENNAN

SARNOFF A. MEDNICK

WILLIAM F. GABRIELLI, Jr.

This chapter examines empirical evidence for a genetic influence in the etiology of antisocial behavior. In this review we will relate results of studies from three approaches to genetic investigation. The first, *family studies,* provides valuable information about increased risk for deviance found among family members of affected individuals. Family studies allow few conclusions about genetic etiology, however, because members of families share environments as well as genes. A second approach, *the study of twins,* offers a somewhat better separation of genetic and environmental effects. The twin studies compare monozygotic (MZ) twins, who are genetically identical, to fraternal, same-sex, dizygotic (DZ) twins who have no more genes in common than other siblings (50%). The research design assumes that the effect of hereditary factors is demonstrated if the MZ twins have more similar outcomes (concordance for deviance) than DZ twins. In almost all studies, the twins are reared together, and the method assumes that the environmental influences upon MZ twins are no different from those upon DZ twins. The possibility exists, however, that environmental influences treat MZ pairs more similarly than DZ pairs. (See the chapter in this volume by Rose for a discussion of twin studies.)

A third approach, *the adoption study,* to a large extent overcomes the possibility of confounding genetic and environmental factors, which limits inferences from the results of twin studies. In this method, the deviant outcomes of adopted children (separated early in life from their biological parents) are compared to the outcomes of their adoptive parents and their biological parents. Similarly in outcome between adoptees and biological parents indicates a genetic effect. In addition, with the application of cross-fostering analysis, relative contributions to deviance from genetic family and from family of rearing may be compared and interactions between genetic and environmental factors may be examined. (See the chapter in this volume by Cadoret for a discussion of adoption studies.)

Family Studies

Family studies of criminality and antisocial behavior have consistently revealed a relationship between parental and offspring criminality (Cloninger & Guze, 1970; Glueck & Glueck, 1974; McCord & McCord, 1958; Roberts, 1978; Robins, 1966; West & Farrington, 1977). In the classic study by Robins (1966), a father's criminal behavior was the single best predictor of antisocial behavior in a child. In terms of genetics, very little can be concluded from such family data alone. The parents have a major influence on the child's environment as well as on his/her genetic make-up; family studies cannot disentangle these hereditary and environmental influences.

Twin Studies

In the first twin-criminality study, the German psychiatrist Lange (1929) found 77% pairwise concordance for criminality for his MZ twins and 12% pairwise concordance for his DZ twins. Lange concluded that "heredity plays a preponderant part among the causes of crime." Subsequent studies of twins (until 1961 there were eight in all) have tended to confirm the direction, but not the extent, of Lange's results. About 60% pairwise concordance has been reported for MZ and about 30% pairwise concordance for DZ twins (see Table 12.1). For a detailed discussion of these twin studies the reader may turn to Christiansen (1977a).

Some of these eight twin studies suffer from the fact that their sampling was rather haphazard. Some were carried out in Germany or Japan during a politically unfortunate period. They reported too high a proportion of MZ twins. That is, all were selected samples from prisons in which concordant MZ pairs are more likely to be brought to the attention of the investigator. Twinship is usually easier to detect in the case of identical twins,

TABLE 12.1.
Twin Studies of Antisocial Behavior—Monozygotic and Same-Sexed Dizygotic Twins Only

Study	Location	Monozygotic			Dizygotic		
		Total pairs	Pairs concordant	% Concordant	Total pairs	Pairs concordant	% Concordant*
Lange '29	Bavaria	13	10	77	17	2	12
Legras '32	Holland	4	4	100	5	1	20
Rosanoff '34	U.S.A.	37	25	68	28	5	18
Stumpfl '36	Germany	18	11	61	19	7	37
Kranz '36	Prussia	32	21	66	43	23	54
Borgstrom '39	Finland	4	3	75	5	2	40
Slater '53	England	2	1	50	10	3	33
Yoshimasu '61	Japan	28	17	61	18	2	11
Total		138	92	67.2	145	45	31.0

SOURCE: Reprinted, by permission, from Mednick et al. (1985). Copyright 1985 by the Academic Press, Inc.
*Pairwise concordance rates.

especially if they end up in the same prison. All of these factors tend to inflate MZ concordance rates in nonsystematic studies.

These twin studies also share another methodological weakness—they all report pairwise concordance rates rather than probandwise concordance rates. Pairwise concordance rates directly reflect the percentage of concordant twin pairs in the ascertained sample. This rate has been shown to be biased in that concordant twin pairs are more likely than discordant twin pairs to be located in a selected sample (Smith, 1974). This bias can be avoided by determining the concordance rate according to the individual rather than the twin pair. The most appropriate measure for this is the probandwise concordance rate, which is equal to the "proportion of cotwins with the trait for individuals independently ascertained" (Smith, 1974: 454).

A national twin study carried out by Christiansen (1977b) minimized the methodological problems discussed above. He studied all twins (3,586 twin pairs) born in a well-defined area of Denmark between 1881 and 1910 and used a national, complete criminality register to ascertain their criminal histories. Probandwise, as well as pairwise, concordance rates were reported. Proband concordance rates were 52% for MZ and 22% for DZ twins. The MZ concordance rate in this unselected twin population was lower than in previous studies. Nevertheless, the MZ rate was 2.3 times greater than the DZ rate; such a difference is consistent with a genetic influence in criminal behavior.

A study by Dalgaard and Kringlen (1976), based on a sample of 139 Norwegian male-male twins, reported 41% probandwise concordance for MZ twins and 25.8% probandwise concordance for DZ twins. Though the differences between MZ and DZ twins were smaller than previous reported studies, they were in the same direction. It is important to note that the Dalgaard-Kringlen twin sample seems to have been drawn relatively heavily from the lower socioeconomic classes, that is, they have a "less-than-normal degree of education, they are to a lesser degree married; and frequency of alcoholism seems higher in this group than in the general population" (Dalgaard & Kringlen, 1976: 221). In Christiansen's larger Danish twin investigation, MZ-DZ concordance differences were considerably lower in subgroups characterized by these sorts of variables. It would seem prudent for Dalgaard and Kringlen to sample additional segments of the

Norwegian twin register in order to attempt to overcome these social class and deviance overrepresentations.

All of the aforementioned twin studies found higher concordance rates for criminality in MZ twins than in DZ twins. This suggests that the greater number of shared characteristics of MZ twins result in greater similarity in criminal behavior than the smaller number of shared characteristics of DZ twins. It is important to note, however, that MZ twins are not only more genetically similar, but also may experience more similar environments than DZ twins.

New methods of statistical modeling have been suggested as useful tools in differentiating the effects of heredity and environment in twin studies of criminality. In a recent twin study of self-reported delinquency, Rowe (1983) utilized a biometrical modeling approach in order to examine the role of genetic, common environmental, and specific environmental factors on the development of delinquent behavior. In this approach the data are fitted to different hypothesized models (using a statistical package called LISREL) in order to determine which models most closely resemble the actual data. Rowe discovered that only those models that contained a genetic component fit well with the data. A purely environmental model containing both common and specific environmental components was rejected.

Rowe's twin study (1983) revealed both the utility and the limitations of current biometrical models in twin research. Although this method allows for a separation of some genetic and environmental factors, it is difficult to conceive of a model that could account for all of the possible genetic and environmental similarities that may exist among twins. Unless all the pertinent variables are included in the model, it is not possible to determine which are more important than others in influencing criminal behavior. Specifically, the omission of certain common environmental factors for MZ twins may result in the premature conclusion that genetic factors play a predominant role in the etiology of criminal behavior.

The use of more appropriate probandwise concordance rates and newer statistical modeling procedures has enhanced the utility of the twin method in genetic research. Nevertheless, some may still question this method due to the possible influence of the greater interpair environmental similarity of MZ twins.

Adoption Studies

The study of adoptions better separates environmental from genetic influences; if adopted children with criminal biological parents were found to commit more crimes than appropriate controls, this would suggest a genetic influence in antisocial behavior. Crowe (1974) found just such a suggestion of genetic influence in an adoption study examining 52 offspring born to incarcerated female offenders in Iowa. Of those 52 adoptees, 7 were convicted as adults in comparison with 1 adoptee conviction in a well-matched control group. Similarly, 6 of the offspring of criminal mothers were diagnosed as having an antisocial personality, in comparison with only 1 adoptee in the control group. Crowe also reported an interaction between genetic and environmental factors such that adoptees who had *both* a criminal mother *and* spent a longer time in temporary placement were found to have the highest rates of conviction.

In a separate Iowa adoption study, Cadoret et al. (1985) compared the rates of alcoholism and antisocial personality in two groups of adoptees: those with antisocial biological family members and those with alcoholic biological family members. The authors reported that approximately one third of the adoptees in each of these groups could be diagnosed with antisocial personality disorder. These rates were significantly higher than the rate of antisocial personality in a matched control group, suggesting a genetic influence in antisocial behavior. It seems likely that the explanation for the high rate of antisocial behavior in adoptees with alcoholic biological families is the fact that many of the alcoholic family members were also antisocial. When this factor was controlled, the authors found that the rate of antisocial personality in this group of adoptees was not significantly elevated over the rate for normal controls.

Cadoret et al. (1987) reported a replication of the above findings with a separate group of adoptees in Iowa. Whereas the first study diagnosed the adoptees through structured personal interviews with the subjects themselves, this second study based the adoptee diagnosis on information gathered only by phone interviews with the adoptive parents. Although the results of the second study replicated the first, the method of data collection in the latter study is of questionable validity.

Sweden

Bohman (1978) also examined the possible genetic influences in alcoholism and criminality in a large adoption study in Sweden. The criminal behavior and alcohol abuse of 2,324 adoptees and their biological parents were assessed through national registers. Results revealed that male adoptees whose biological fathers were registered for criminal behavior alone (no alcohol abuse) were not more criminal than male adoptees whose biological fathers had no record of criminal behavior (12.5% vs 12.0%). As a result of this finding, Bohman concluded that there is no genetic influence in criminality. This conclusion, however, is based on a limited view of the availability data. Bohman was careful to separate out the alcohol abuse and the criminality in the biological fathers in this analysis. He did not, however, distinguish between registrations for criminality alone and criminality plus alcohol abuse in the sons. If one separates these two groups, it is discovered that there are significantly more "criminal only" sons of "criminal only" biological fathers than there are "criminal only" sons of other fathers in the study (8.9% vs 4.9%, $p < .05$).

In a later study of the Swedish adoption cohort, Bohman et al. (1982) recognized and expanded on this distinction between adoptees who were registered for criminality only and those registered for both criminality and alcohol abuse. They discovered that the adoptees who were registered for criminality only were likely to have committed property crimes, whereas those who were registered for both criminality and alcohol abuse were more likely to have committed more serious, violent crimes. It is interesting to note that the genetic influence was found to be significant in the case of the property criminals, but *not* in the case of the alcoholic or violent criminals.

Further studies of the Swedish cohort (Cloninger et al., 1982; Sigvardsson et al., 1982) have revealed important findings about sex differences in the genetic transmission of property crime. It is widely know that men have more social and environmental pressures toward criminal behavior than women. It is reasonable to hypothesize, then, that those females who do become antisocial may have a stronger genetic predisposition toward this behavior than the males who become antisocial. This hypothesis was supported by Sigvardsson et al. (1982) who found that the female adoptee property criminals in their cohort had a much higher percentage of biological parents who were property criminals than did the males (50% vs 21%, $p < .05$).

Denmark

A register of all nonfamilial adoptions in Denmark in the years 1924–1947 has been established in Copenhagen at the Psykologisk Institute by a group of American and Danish investigators headed by Kety et al. (1968). There are 14,427 adoptions recorded, including information on the adoptee and his or her biological and adoptive parents. Two completed investigations utilized information from this Danish adoption register. In the first, Schulsinger (1972) selected 57 psychopaths from psychiatric registers and police files and compared them to 57 nonpsychopath control adoptees matched for sex, age, social class, neighborhood of rearing, and age of transfer to the adoptive family. He found that 5 of the psychopathic adoptees had psychopathic biological fathers. In contrast only 1 of the control adoptees had a psychopathic biological father. Although the numbers are small, the direction of the findings supports the hypothesis of heredity as an etiological factor in psychopathy.

Using the same adoption register, Mednick et al. (1984) compared the court conviction histories of all 14,427 adoptees, their biological mothers and fathers, and their adoptive mothers and fathers. The conviction rates of the completely identified members of this cohort are shown in Table 12.2. The rates for biological fathers and their adopted-away sons were considerably higher than those of adoptive fathers. Most of the criminal adoptive fathers were one-time offenders, whereas male adoptees and their biological fathers were more heavily recidivistic. The conviction rates of the women in this study were lower than those of the men but followed the same pattern. In view of the low conviction rates for women, the analyses in the study concentrated on male adoptees.

The size of this population permitted the segregation of subgroups of adoptees with certain combinations of convicted and nonconvicted biological and adoptive parents in a design analogous to the cross-fostering model used in behavior genetics. If neither the biological nor the adoptive parents were convicted, 13.5% of the sons were convicted. If the adoptive parents were

TABLE 12.2.
Conviction Rates of Completely Identified Members
of Adoptee Families

Family member	Number identified	Number not identified	Conviction Rate by Number of Convictions			
			0	1	2	>2
Male adoptees	6,129	571	0.841	0.088	0.029	0.049
Female adoptees	7,065	662	0.972	0.020	0.005	0.003
Adoptive fathers	13,918	509	0.938	0.046	0.008	0.008
Adoptive mothers	14,267	160	0.981	0.015	0.002	0.002
Biologic fathers	10,604	3,823	0.714	0.129	0.056	0.102
Biologic mothers	12,300	2,127	0.911	0.064	0.012	0.013

Source: Reprinted, by permission, from Mednick et al. (1985). Copyright 1985 by the Academic Press, Inc.

convicted and the biological parents were not, this figure rose to only 14.7%. However, if the adoptive parents were not convicted and the biological parents were, 20.0% of the sons were convicted. If the adoptive parents as well as the biological parents were convicted, 24.5% of the sons were convicted. These data favor the assumption of a partial genetic etiology.

It is important to note that Mednick et al. (1984) found that these significant relationships between biological parent convictions and adoptee son convictions existed only for property offenses, not for violent offenses. These results paralleled those of the Swedish adoption studies that found a significant genetic influence for property crimes rather than for violent crimes or alcohol abuse (Bohman et al., 1982; Cloninger et al., 1982).

The Mednick et al. (1984) study further replicated the Swedish findings on sex differences. The Danish adoption study revealed that the relation between biological *mother* convictions and adoptee convictions was significantly stronger than that between biological *father* convictions and adoptee convictions. A recent examination of the data (Baker et al., 1989) more closely replicated the Swedish finding (Sigvardsson et al., 1982), in that female criminal adoptees were found to have a higher percentage of criminal biological parents than male criminal adoptees.

Perhaps the most important finding of Mednick et al. (1984) is

TABLE 12.3.
Proportion of Chronic Offenders (3 or More Convictions),
Other Offenders (1 or 2 Convictions), and Nonoffenders
in Male Adoptees as a Function of Number of Convictions
in the Biological Parents

| Male adoptee | Biological parent convictions | | | |
convictions	0	1	2	3 or more
None	0.87	0.84	0.80	0.75
1 or 2	0.10	0.12	0.15	0.17
3 or more	0.03	0.04	0.05	0.09
Number of adoptees	2492	574	233	419

Source: Reprinted, by permission, from Mednick et al. (1985). Copyright 1985 by the Academic Press, Inc.
Note: Cases in which adoptive parents have been convicted of criminal law violations are excluded.

one that has not been reported in other adoption studies; that finding concerns genetic influences on recidivism. Table 12.3 shows how chronic offenders, other offenders (one or two convictions), and nonoffenders were distributed as a function of amount of crime in biological parents. Note that the proportion of chronic adoptee offenders increased as a function of recidivism in the biological parents.

Another way of expressing this concentration of crime is that the chronic male adoptee offenders with biological parents having three or more offenses numbered only 37. They made up 1% of the 3,718 adoptees in Table 12.3, but they were responsible for 30% of the total male adoptee convictions. The mean number of convictions for the chronic adoptee increased sharply as a function of biological parent recidivism. These results suggest that not only is a genetic factor more important for property crimes than for violent ones, but it also plays a very significant role in repeat offending.

Special Considerations in Adoption Studies

Family and twin studies are often criticized because the criminal subjects and their relatives almost always share both genes and environment. This criticism is less justified in the case of adoption studies, which seem to have greater face validity. Because of this and because not everyone is equally familiar with the circumstances of adoption, it may be useful to consider the problems in using adoptions as a research tool. We will do so in the context of the aforementioned Danish adoption research on criminality.

If anything were to be found that lessened the independence of genetic and environmental influences in the adoption studies, it would serve to temper the impact of these findings. Hutchings (1972) has pointed out that the major adoptive agency in Denmark had a policy of attempting to match the biological and adoptive families for vaguely defined social characteristics. A significant, but modest, correlation of the social class of biological and adoptive fathers ($r = .22, p < .001$) attests to their success. The variable of social class has been found, on its own, to be significantly related to criminal behavior in adoptive offspring (Van Dusen et al., 1983). To what extent, then, does the correlation between the biological fathers' and the sons' social class of rearing explain the "genetic" findings? Stepwise multiple regression analyses, with adoptee criminality as the dependent variable and the father's criminality and social class as the independent variables, indicated that independent of socioeconomic status (SES), biological parent criminality is significantly related to adoptee criminality. Moreover, it was found that the relation between biological parent and adoptee criminal convictions existed at *each level* of adoptive parent SES (Mednick et al., 1984). Social class influences apparently do not explain away the genetic effect of a biological parent's criminality.

The amounts of environmental and genetic variability in a study population should be considered in terms of the generalizability of its results. In a study population with relatively little environmental variability, the genetic effects are enhanced. One might suggest that the genetic influences found in the Danish adoption study are enhanced for that reason, and thus should not be generalized to other populations. It is true that relatively low environmental variability exists in Denmark. It is also true, however, that relatively low *genetic* variability exists in Denmark, and that this is likely to *reduce* the effects of genetic influence. It is suggested that the effects of low environmental and low genetic variability in Denmark may work to cancel each other out.

The screening of the biological and adoptive parents and subsequent matching of adoptive parents to adoptee by adoption agencies can produce an adoptee population with characteristics that are very different from the normal population. Adoptive parents, for example, may tend toward higher SES and be more educated than parents in general. Any such unusual representation in the population of adoptees must be considered in the application of findings to nonadoptee populations.

Another potential problem in using an adoptee population to study genetic influences of criminal behavior is that at the time of the adoption the adoptive parents are often informed by the adoption agency of deviance in the biological family. Such a disclosure could conceivably result in labeling of the adoptee's behavior in accordance with this information. This labeling might, in turn, influence the probability of the adoptee manifesting deviance. Such an indirect relationship between the biological parents' behavior and adoptee behavior could confound the findings. Reading some of the old adoption files in the Danish study makes it clear that serious deviance in the biological parents was more or less routinely reported to the prospective parents unless they refused the information.

As a test of the influence of such a disclosure about biological family history of deviant behavior upon the findings reported above, families in which the adoptive parents might have known about the biological family criminal history were examined apart from those in which the adoptive families did not know of any deviance. If the biological father's criminal career began at some time after the birth of the child, this information could not have been transmitted to the adoptive parents. On the other hand, in cases in which the criminal offending of the biological father started before the birth of the adoptee, the information probably was communicated. Of the convicted biological parents, 37% had their first registration for criminality before the birth of the child and 63% began their criminal career after the birth of the adoptee. In the cases where the biological parent committed his/her first offense before the birth of the adoptee, 19.8% of the male adoptees became criminal. In the cases where the biological parent committed his/her offense after the birth of the adoptee, 20.4% of the adoptees became criminal. Our tentative conclusion from this analysis is that the possibility of the adoptive family being informed of the biological parents' criminality is not related to an altered likelihood that the adoptive son will become criminal.

In considering the results of adoption studies one must keep in mind potential problems such as limited generalizability, selective placement, and parental labeling of the child. Our consideration of these potential problems in the Danish adoption study points out the need to attend to them as potential confounds. Nevertheless, even when many of these potential pitfalls were controlled, the results of the Danish adoption study remained significant. This supports the hypothesis of a genetic influence in criminal behavior.

Conclusion

This review has demonstrated that genetic factors can and do influence certain types of criminal behavior. What are the implications of this knowledge? First, we must add biological factors to our list of causes of crime; it is through heritable biological structures and processes that the genes exert their influence. Second, we must try to identify the specific biological mechanisms through which heritable predispositions toward criminal behavior are expressed. By identifying these mechanisms we can learn how to successfully treat and prevent criminal behavior.

Our work in Denmark suggests that one biological mechanisms involved in a predisposition toward criminal offending may be responsiveness and recovery of the autonomic nervous system (ANS). We know that slow ANS responding is characteristic of adult criminal offenders (Mednick & Volavka, 1980). Low responsiveness has also predicted delinquency and recidivism over a 10-year period in a Danish longitudinal study (Loeb & Mednick, 1977). We have noted this ANS characteristic in the children of criminal offenders (Mednick, 1977), and in a twin study we have found that ANS responsiveness is partially genetically determined (Bell et al., 1977).

Mednick (1977) suggests that ANS responsiveness may play a role in the social learning of law-abiding behavior and theorizes that faster ANS recovery should be associated with greater reinforcement and increased learning of the inhibition of antisocial responses. Slow ANS recovery, on the other hand, should be associated with poor learning of the inhibition of antisocial responses. This theoretical model illustrates how a biological variable (that is partially genetically determined) may play a role in the social learning of criminal behavior. This may be seen as an example of the integration of genetic, biological, and social learning mechanisms in the study of the etiology of criminal behavior. It is hoped that complete integrations such as these can help to suggest appropriate treatment and preventive interventions for antisocial behavior.

Acknowledgment

Preparation of this chapter was supported by Grant 87-IJ-CX-0063 and a Graduate Fellowship from the National Institute of Justice.

References

Baker, L., Mack, W., Moffitt, T. & Mednick, S. (1989) Sex differences in property crime in a Danish adoption cohort. *Behavior Genetics* 19(3):355–370.

Bell, B., Mednick, S. A., Gottesmann, I. I. & Sergeant, J. (1977) Electrodermal parameters in young, normal male twins. In *Biosocial Bases of Criminal Behavior*, ed. S. A. Mednick & K. O. Christiansen, pp. 217–226. New York: Gardner.

Bohman, M. (1978) Some genetic aspects of alcoholism and criminality. *Archives of General Psychiatry* 35:269–276.

Bohman, M., Cloninger, C. R., Sigvardsson, S. & von Knorring, A. L. (1982) Predisposition to petty criminality in Swedish adoptees. I. Genetic and environmental heterogeneity. *Archives of General Psychiatry* 39:1233–1241.

Borgstrom, C. A. (1939) Eine Serie von Kriminellen Zwillingen. *Archiv fur Rassenbiologie.*

Cadoret, R. J., O'Gorman, T. W., Troughton, E. & Heywood, E. (1985) Alcoholism and antisocial personality: Interrelationships, genetic and environmental factors. *Archives of General Psychiatry* 42:161–167.

Cadoret, R. J., Troughton, E. & O'Gorman, T. W. (1987) Genetic and environmental factors in alcohol abuse and antisocial personality. *Journal of Studies on Alcohol* 48:1–8.

Christiansen, K. O. (1977a) A review of studies of criminality among twins. In *Biosocial Bases of Criminal Behavior*, ed. S. A. Mednick & K. O. Christiansen, pp. 45–88. New York: Gardner.

Christiansen, K. O. (1977b) A preliminary study of criminality among twins. In *Biosocial Bases of Criminal Behavior* ed. S.A. Mednick & K. O. Christiansen, pp. 88–108. New York: Gardner.

Cloninger, C. R. & Guze, S. B. (1970) Female criminals: Their personal, familial, and social backgrounds. *Archives of General Psychiatry* 23:554–558.

Cloninger, C. R., Sigvardsson, S., Bohman, M. & von Knorring, A. L. (1982) Predisposition to petty criminality in Swedish adoptees: II. Cross-fostering analysis of gene-environment interaction. *Archives of General Psychiatry* 39:1242–1247.

Crowe, R. R. (1974) An adoption study of antisocial personality. *Archives of General Psychiatry* 31:785–791.

Dalgaard, O. S. & Kringlen, E. (1976) A Norwegian twin study of criminality. *British Journal of Criminology* 16:213–232.

Glueck, S. & Glueck, E. (1974) *Of Delinquency and Crime.* Springfield, IL: Charles C. Thomas.

Hutchings, B. (1972) *Genetic and Environmental Factors in Psychopathology and Criminality.* M. Phil. Thesis, University of London.

Kety, S. S., Rosenthal, D., Wender, P. H., & Schulsinger, F. (1968) The types and prevalence of mental illness in the biological and adoptive families of adopted schizophrenics. In *The Transmission of Schizophrenia* ed. D. Rosenthal & S. S. Kety. Oxford: Pergamon.

Kranz, H. (1936) *Lebensschicksale Kriminellen Zwillinge.* Berlin: Julius Springer.

Lange, J. (1929) *Verbrechen als Schiskal.* Leipzig: George Thieme. (English edition, London: Unwin Brother.)

Legras, A. M. (1932) *Psychese en Criminalitet bij Twellingen.* Utrecht: Kemink en Zonn.

Loeb, J. & Mednick, S. A. (1977) A prospective study of predictors of criminality: Electrodermal response patterns. In *Biosocial Bases of Criminal Behavior* ed. S. A. Mednick & K. O. Christiansen, pp. 245–254. New York: Gardner.

McCord, J. & McCord, W. (1958) The effects of parental role model on criminality. *Journal of Social Issues* 14:66–75.

Mednick, S. A. (1977) A biosocial theory of the learning of law-abiding behavior. In *Biosocial Bases of Criminal Behavior,* ed. S. A. Mednick & K. O. Christiansen, pp. 1–8. New York: Gardner.

Mednick, S. A., Gabrielli, W. F. & Hutchings, B. (1984) Genetic influences in criminal convictions: Evidence from an adoption cohort. *Science* 891–894.

Mednick, S. A., Moffitt, T., Gabrielli, W. F. & Hutchings, B. (1985) Genetic factors in criminal behavior: A review. In *Development of Antisocial and Prosocial Behavior,* ed. D. Olweus. New York: Academic Press.

Mednick, S. A. & Volavka, J. (1980) Biology and crime. In *Crime and Justice: An Annual Review of the Research,* ed. N. Morris & M. Tonry, pp. 85–158. Chicago: University of Chicago Press.

Robert, L. N. (1978) Aetiological implications in studies of childhood histories relating to antisocial personality. In *Psychopathic Behavior:*

Approaches to Research, ed. R. D. Hare & D. Schalling, pp. 225–271. New York: John Wiley.

Robins, L. N. (1966) *Deviant Children Grown Up*. Baltimore: Williams & Wilkins.

Rosanoff, A. J., Handy, L. M. & Rosanoff, F. A. (1934) Criminality and delinquency in twins. *Journal of Criminal Law and Criminology* 24:923–934.

Rowe, D. C. (1983) Biometrical genetic models of self reported delinquent behavior: A twin study. *Behavior Genetics* 13:473–489.

Schulsinger, F. (1972) Psychopathy: Heredity and environment. *International Journal of Mental Health* 1:190–206.

Sigvardsson, S., Cloninger, C. R., Bohman, M. & von Knorring, A. L. (1982) Predisposition to petty criminality in Swedish adoptees: III. Sex differences and validation of the male typology. *Archives of General Psychiatry* 39:1248–1253.

Slater, E. (1953) The incidence of mental disorder. *Annals of Eugenics* 6:172.

Smith, C. (1974) Concordance in twins: Methods and interpretation. *American Journal of Human Genetics* 26:454–466.

Stumpfl, F. (1936) *Die Ursprunge des Verberchens. Dargestellt am Lebenslauf von Zwillingen*. Leipzig: Georg Thieme.

Van Dusen, K. T., Mednick, S. A., Gabrielli, W. F. & Hutchings, B. (1983) Social class and crime in an adoption court. *The Journal of Criminal Law and Criminology* 74:249–269.

West, D. J. & Farrington, D. P. (1977) *The Delinquent Way of Life*. New York: Crane Russak.

Yoshimasu, S. (1961) The criminological significance of the family in light of the studies of criminal twins. *Acta Criminologiae et Medicinae Legalis Japanica* 27.

Chapter 13
Psychiatric Investigations and Findings from the Minnesota Study of Twins Reared Apart

NANCY L. SEGAL
WILLIAM M. GROVE
THOMAS J. BOUCHARD, Jr.

Introduction

The Minnesota Study of Twins Reared Apart was launched in 1979 by investigators in the Departments of Psychology and Psychiatry, at the University of Minnesota. Monozygotic (MZA) and dizygotic (DZA) twins reared apart are brought to Minneapolis for extended study. Participants remain in Minneapolis for an entire week, with the total evaluation lasting approximately 50 hours. Spouses or other close relatives accompany the twins to the University of Minnesota, and complete an abbreviated version of the psychological test battery. A sample assessment schedule appears in Figure 13.1. The major research objectives of the project are to estimate genetic and environmental influences on a wide variety of medical and psychological traits. The correlations between the twins estimate the magnitude of genetic influence on each trait. The correlations between early medical and social life history differences, and current medical and psychological differences, between the twins estimate specific environmental influences.

Twins are located primarily through media publicity contacts, through various adoption groups, searches carried out by the research team, and direct contact from the twins or their relatives. Unlike all previous studies of reared-apart twins, particularly

	SUNDAY	MONDAY	TUESDAY	WEDNESDAY	THURSDAY	FRIDAY	SATURDAY
07:00				Blood Samples			
08:00		Respirator / Cardiology	Wechsler Adult Intelligence Scale	Breakfast	Mental Abilities	Psychomotor	Mental Abilities
09:00		Chest X-ray	Stress EKG #1	Heart Monitors	Psychomotor	Information Processing / Life Stress Int.	Psychomotor
10:00		Echo/Vector Cardiogram	Stress EKG #2	Blood Samples Allergy Interview	Mental Abilities	Life Stress Int. / Information Processing	Info. Proc. / Inventories — Inventories / Info. Proc.
11:00			Inventories	Periodontal/Dental Exams	Vitalogs / Inventories		
12:00		Lunch	Lunch	Lunch	Lunch	Lunch	Lunch
13:00							

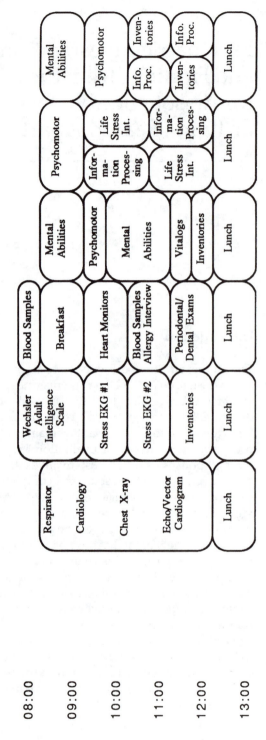

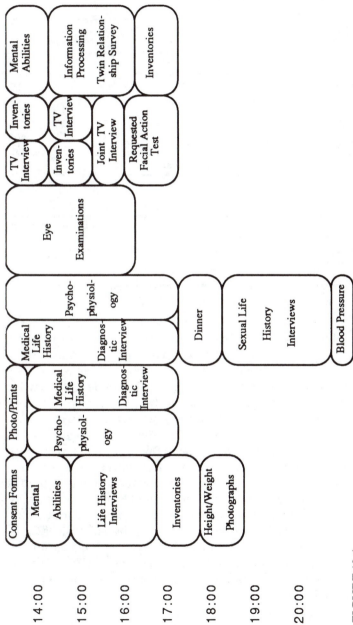

FIGURE 13.1
Assessment schedule for participants in the Minnesota Study of Twins Reared Apart.

TABLE 13.1.
Mean Age at Testing, Separation, and Reunion of Twins
Reared Apart from Five Studies

Study	N (Pairs)	Mean age at:		
		Testing (years)	*Separation* (years)[a]	*Reunion* (years)
MZA twin pairs				
Newman et al. (1937)	19	26.1	1.3	12.1
Shields (1962)[b]	38	39.5	1.3	11.1
Juel-Nielsen (1965)	12	51.4	1.4	15.1
Bouchard et al.	49[c]	38.9	0.3	30.3
DZA twin pairs				
Pedersen et al. (1985)	34–38	59.0	1–10	85% after 17
Bouchard et al.	25[d]	40.8	1.1	37.2

[a]In the first three Studies, age at reunion is often estimated. If it was reported that the twins met in childhood, an age of 6 years is assumed.
[b]Descriptive data are based on cases for which IQ measures are available. The full sample consisted of 44 cases.
[c]Includes 2 MZA triplet sets; 59% of twin pairs are female.
[d]64% of twin pairs are female; 16% of twin pairs are opposite sex.

those of Newman et al. (1937) and Shields (1962), the Minnesota Study of Twins Reared Apart makes every attempt to pursuade all located twins (including DZ twin pairs and more recently opposite-sex twin pairs) to participate. This sample of MZA twins is, therefore, somewhat superior to those in previous studies because selection for participation is not based on twin similarity. In addition, the sample of DZA twins provides an informative control for placement bias. (Psychological and medical studies of both reared-apart MZ and DZ twins are now ongoing in Sweden (Plomin et al., 1988) and in Finland (Langinvainio et al., 1984). Compared with previous samples, twins in the Minnesota sample were separated much earlier, reunited much later, and (with the exception of twins studied by Shields, 1962) assessed at an older age. Additional details concerning recruitment of participants are presented in Bouchard et al. (1990). Sample characteristics for the different studies of reared-apart twins are compared in Table 13.1.

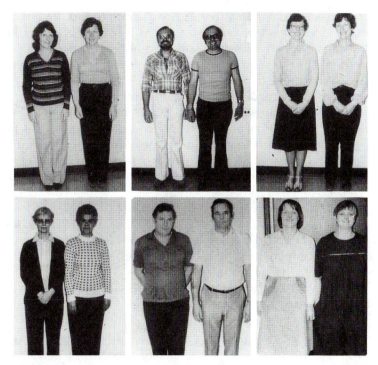

FIGURE 13.2
MZA (top row) and DZA (bottom row) twin pairs. Photo
courtesy of Dr. Thomas J. Bouchard, Jr., Director, Minnesota
Center for Twin and Adoption Research.

Zygosity was determined by analyses of eight blood group
systems, four serum proteins, six red blood cell enzymes, fingerprint ridgecount, ponderal index, and cephalic index. The
probability of misdetermining zygosity by this procedure is less
than .001 (Lykken, 1978). Several MZA and DZA twin pairs who
have participated in the study are shown in Figure 13.2.

The Reared-Apart Twin Design

The study of twins reared apart is a variant of the classic twin
method. In the classic twin method, resemblance within MZ and

DZ twins is compared with reference to a behavior or trait of interest. Greater resemblance within MZ twinships than DZ twinships is consistent with a genetic contribution to the trait in question. The classic twin method assumes equally correlated environments for the two types of twins. This assumption has been especially challenged by critics of the twin method, although there is considerable evidence to refute these objections. (See Bouchard, 1984, for a comprehensive review.) The key advantage of the reared-apart twin design over the ordinary twin design is that the study of separated MZ twins provides a direct estimate of heritability, given that the twins have been placed in homes uncorrelated for environmental factors casually related to the trait in question. In addition, this design results in greatly reduced sampling variability for estimates of genetic model parameters; for example, estimates of heritability based on MZA twins have one-fourth the sampling variability of estimates based on MZ-DZ twin comparisons (Lykken et al., 1982).

A discussion of placement effects as a possible source of bias in studies of reared apart twins is informative:

The requirement of randomization of environments in MZA twin designs means exactly that—randomization of cases across trait-relevant environments. It does not mean or imply that it is not possible to find members of the same pair in highly similar environments. Randomization requires that there be such cases, else they would not be randomized. Sims (1931) matched a large group of children on age alone and obtained a correlation of .04. An examination of these pairs would undoubtedly show that some were fully matched on factors similar to those described by Taylor. Environmental effects will be detected by correlating specific environmental differences with twin differences. Newman et al. carried out such an analysis and found some sizeable correlations. The important point is that these correlations explain the variation among twins not already accounted for by heredity. The less-than-perfect correlation between twins' IQs already reflects these effects. (Bouchard, 1983:181–182)

There are two sampling schemes commonly used for recruiting twins for studies of psychiatric symptoms and illnesses. The most common scheme, used for example at the Twin Registry of the Maudsley Institute of Psychiatry, in London, England, asks psychiatric patients if they are a twin and if so, subsequently investigates them and their co-twin. The second method, used in the Minnesota Study of Twins Reared Apart, samples twins from the general population. The advantages of the latter procedure are that, first, a reasonable number of MZA twins can be found; this could not be accomplished by screening only psychiatric patients unless a very large number of medical facilities were involved over many years. Second, this method avoids a number of potentially important sampling biases. Bias in case detection can occur because symptom status is only one among many variables determining who seeks or who is referred for treatment. Reluctance to admit problems, inability to pay, and lack of local services can all militate against a case coming to attention. The present quasi-epidemiological design resolves this problem, since whether or not twins have sought treatment for a given nervous, emotional, or behavioral problem, they will presumably be about equally likely to be recruited into our sample.

The disadvantage of the MZA design is the obverse side of the advantage just discussed. The relative health of the twins means that certain less common disorders (e.g., schizophrenia) are very difficult to study with this method. Furthermore, cases identified in normal proband studies are often less severe than those seen in medical facilities. For this reason, in the present study, attention has so far been restricted to conditions that are relatively common in the general population. In view of the foregoing, summary symptom-count scores, rather than diagnostic labels, have been analyzed because these scores show more analyzable variability.

Psychiatric Assessment Methods

Members of a family (twin pair or triplet set) are assessed during the same week whenever possible. Measures related to psychopathology are the Minnesota Multiphasic Personality Inventory (MMPI), a Fear Survey, the General Behavior Inventory (GBI), a comprehensive structured Medical Life History Interview, a Sexual Life History assessment, and the Diagnostic Interview Schedule (DIS; Robins et al., 1981). Only the MMPI and Medical Life History Interview were in place at the beginning of the study. Brief descriptions of these instruments are provided below.

Minnesota Multiphasic Personality Inventory

The MMPI (Hathaway & McKinley, 1943) is a widely used instrument for personality assessment that includes more than 500 items concerning behavioral style and preference; it requires responses of "true," "false," or "can't say." Response analysis yields nine clinical scales (e.g., hypochondriais, depression). Twins complete the MMPI independently on the same day. The scale data may ultimately prove informative in their own right, as well as potentially provide an exploration of personality and symptom factors related to psychiatric disorders diagnosed from interviews. Analyses of psychiatric abnormalities based upon the MMPI have not, as yet, been completed.

Fear Survey

Developed for research purposes by Gregory Carey in 1977, the Fear Survey, unlike many other surveys, includes more than a single item for each fear. The items are organized into seven content scales: normal animals, nasty animals, nature fears, orderliness, health fears, agoraphobic-like fears, and social fears. These seven broad scales are each composed of more specific fears. The nature fear scale, for example, includes items relevant to fears of storms, water, and heights. Also, two validity scales—one for physically dangerous events and one for stressful social situations—are included to tap response tendencies toward high or low ratings of all items, and high or low ratings of unpleasant items, respectively. The Fear Survey includes 156 items, each rated on a Likert scale from 0 (no fear) to 12 (frozen with fear).

General Behavior Inventory

The GBI is a self-report psychiatric inventory that is independently completed by each twin. The construction of this inventory was based on a binomial model of psychiatric disorder, for the purpose of separating a population into cases and noncases. The inventory score provides a probability estimate of individuals' classification into one of these categories. Scoring yields a probability estimate that the respondent can be classified as a psychiatric case. Both intensity and duration of a particular symptom or behavior are requested, as demonstrated by the following sample item: Have you been accident prone? 1 (never, or hardly ever), 2 (sometimes), 3 (often), or 4 (very often). Additional information concerning the construction and application of the GBI is available in DePue et al., (1981).

Medical History Assessment

During individual interviews, symptoms and histories of illnesses, injuries, and surgeries are reviewed. In addition, twins undergo a general physical examination, cardiovascular examination (chest X-ray, vector cardiogram, EKG, 24-hour EKG using a holter monitor, overnight blood pressure monitor [Hanson et al., 1984], and stress EKG [Hanson et al., 1989]), pulmonary function examination (Hankins et al., 1982), allergy testing (history of exposure to allergens, IgE, serological tests for allergens [RAST; Kohler et al., 1985]), basic laboratory tests (hematological blood count, renal chemistry battery), ophthalmological examination (visual acuity, refraction, ocular pressure, eye dilation, and retinal photographs, and fundus eye examination [Knobloch et al., 1985]), dental and periodontal examinations (complete intra- and extra-oral-facial exam, evaluation for missing teeth, radiographic survey, ratings of attachment loss, and dental calculus [cf. Boraas et al., 1988]), psychophysiology examinations (EEG; Lykken, 1982; Stassen et al., 1988), tests of skin conductance (Lykken et al., 1988), and collection of previous medical records.

Sexual Life History Assessment

This includes a Sexual Meaning Survey, Sexual Life History Timeline, and Sexual Events Questionnaire (of which there are separate versions for males and for females), followed by a structured interview. The various components of this assessment were established by Joseph Bohlen, now at the University of Southern Illinois Medical Center, and colleagues. Trained research associates and advanced graduate students currently conduct the interviews.

The Sexual Meaning Survey presents a series of word pairs (organized as a semantic differential) to determine the sexual meaning or significance of these concepts (e.g., weak-strong; duty-pleasure) for each individual. The Sexual Life History Timeline requests information about the occurrence and timing of developmental events and other experiences associated with sexual development (e.g., age at first awareness of gender differences, age at first intercourse). The Sexual Events Questionnaire obtains information about past and current sexual attitudes and practices (e.g., attitudes and practices regarding contraception, frequency of intercourse). The structured interview enables twins to elaborate upon their developmental sexual history, premarital history, and marital history.

Diagnostic Interview Schedule

The DIS is administered to twins on consecutive days by different interviewers (in all but a very few cases). Interviewers, of course, never confer before completing all ratings. Twins are specifically enjoined not to discuss the content of their interviews with each other. The DIS was first introduced as a standard part of the assessment procedure in March, 1982. Prior to that time, 42 twin pairs had been assessed using an unstructured psychiatric interview administered to each twin separately by the same team of investigators. Currently, a longitudinal follow-up study has begun, enabling repeated administration of the DIS to twins, as well as initial administration of the DIS to returning twins who had not received it previously.

The DIS covers major diagnoses in DSM-III (American Psychiatric Association, 1980), including the following: schizophrenia, schizoaffective disorder, mania, major depression, dysthymia, substance abuse disorders (alcohol abuse and dependence and abuse and dependence of a number of other psychoactive substances), anorexia nervosa and bulimia, antisocial personality, somatization disorder, compulsive gambling, sexual dysfunction, homosexuality, and anxiety disorders (panic, phobic, generalized anxiety and obsessive-compulsive).

Overview of the Findings

A considerable body of psychiatric and medical data are now available on the reared apart twins. Many of these data have not, as yet, been analyzed. A sampling of findings from three different phases of the psychiatric investigation (DIS, case study of Tourette syndrome, and Sexual Life History) are presented below. An early paper reporting preliminary findings of psychiatric disturbances is available (Eckert et al., 1981).

Substance Abuse and Antisocial Personality

Data on 32 twin pairs (31 twin pairs plus 1 set of triplets) were recently analyzed quantitatively (Grove et al., 1990). In order to produce scores with more interindividual variation than diagnoses, signs counting toward Research Diagnostic Criteria (RDC; Spitzer et al., 1978), St. Louis group or Feighner Criteria (Feighner et al., 1972), and DSM-III diagnoses of psychoactive substance use disorders and antisocial personality were tallied. Items not covered (due to DIS rules that requires skipping some items if preliminary questions are answered in the negative) were counted as absent.

Summary scores showing useful individual differences were grouped into four content areas: childhood antisocial behaviors (those manifested before age 15 years), adult antisocial behaviors (manifest since age 15 years), alcohol abuse or dependence items, and drug abuse items. Examples of childhood antisocial behaviors include playing hooky, running away from home overnight at least twice, stealing, repeated fighting, and juvenile court involvements. Examples of adult antisocial behaviors include neglect of children, maltreatment of spouse, felony convictions, gross job instability, and repeated divorce. Examples of items related to alcohol difficulties include blackouts, drinking before breakfast, job, school, family, or legal difficulties associated with drinking, and inability to control drinking. Drug items include craving, inability to cut down or stop, tolerance, and withdrawal symptoms. Since signs of DSM-III alcohol or drug dependence (medical complications, withdrawal symptoms, tolerance) were rare in this nonclinical sample, dependence items were pooled with abuse items. The resulting four scales (alcohol, drug, child antisocial, and adult antisocial) had coefficient alpha reliabilities of .91, .89, .71, and .62. Given that the scales had markedly positively skewed distributions, they were transformed to approximate normality using Blom rank scores.

It is well documented that alcohol and drug problems go hand in hand with antisocial behavior (Guze et al., 1969). Moreover, longitudinal data also suggest considerable stability of antisocial behavior from childhood into adulthood (Robins, 1966). Such correlations could have either environmental or genetic explanations, or both. For example, antisocial behavior in childhood could predict similar behavior in adulthood because both behavioral patterns are consequences of adverse child-rearing environments. On the other hand, it is conceivable that these two problems co-occur because they reflect common heritable core dispositions (e.g., impulsivity coupled with nonconformity). The data were therefore examined to discover whether genetic factors underlying

TABLE 13.2.
Phenotypic Correlations, Heritabilities,
and Genetic Correlations (± Standard Error)

Score	Alcohol	Drug	Child	Adult
Alcohol	.13 ± .030	.258	.27	.28
Drug	.78 ± .027	.46 ± .019	.48	.62
Child	.54 ± .053	.87 ± .008	.42 ± .021	.47
Adult	.75 ± .044	.53 ± .031	.62 ± .030	.29 ± .026

Source: Reprinted, by permission, from Grove et al. (1990). Copyright 1990 by the Society of Biological Psychiatry.

one trait overlapped those underlying another. A useful way to do this is by computing genetic correlations. These statistics, under certain assumptions, estimate the degree to which two traits share common genetic causes (Carey, 1988). Table 13.2 presents heritabilities and genetic correlations, along with phenotypic correlations, for the four scales. All parameters were estimated by maximum likelihood. Phenotypic correlations appear above the diagonal, heritabilities appear on the diagonal, and genetic correlations appear below the diagonal.

The data show that all traits are heritable, although notably less so than traits such as IQ. There is substantial genetic correlation between all traits. In particular, child and adult antisocial behaviors correlate genetically .62, which supports the DSM-III convention of considering both to be signs of the same disorder. Moreover, alcohol and drug problems also genetically correlate. This suggests that in some kinds of family studies (e.g., segregation and linkage analysis), one might wish to count individuals with clear-cut manifestations of any of these categories of behavior as affected. It is possible that specific genetic dispositions exist, leading to dependence on specific psychoactive substances. In a nonclinical sample, such as the present one, however, it is possible that carriers for genetic predispositions to preferentially abuse particular substances might be infrequently recruited. The data are consistent either with a general liability to substance abuse or liabilities for abusing specific substances, in addition to a general liability.

Case Study of Tourette Syndrome

Research interest in genetic and environmental influences on Tourette syndrome (TS) has risen substantially within recent years. The majority of researchers acknowledge that genetic influences play a role in the onset and progression of this disorder, although varied environmental influences may modify expression.

Price et al. (1985) presented the only systematically conducted twin study of TS. Concordance figures of 53% for MZ twin pairs and 8% for DZ twin pairs were reported. These figures increased to 77% for MZ twin pairs and 23% for DZ twin pairs when the diagnostic criteria included "any tics" in the co-twins. These findings were interpreted as providing support for the role of genetic factors in the development of TS. Twins in this study were raised together, however, so the possibility of imitation of TS-related behaviors cannot be completely eliminated.

Detailed case histories provide informative means for examining and generating hypotheses relevant to specific issues. A set of reared-apart triplets, consisting of an MZ female twin pair and DZ male co-triplet, were recruited into the Minnesota Study of Twins Reared Apart in 1981, at the age of 50 years. Separate adoptions occurred at approximately 2 months of age, and initial contact did not take place until the age of 47 years. Additional details concerning the reunion and separation of these triplets are provided in Segal et al. (1990a).

In addition to completing the standard medical and psychological assessment battery administered to reared-apart twins, the triplets completed a Tourette syndrome survey, which was largely based on questionnaires developed at Yale University by David L. Pauls and Kenneth K. Kidd. These protocols were administered to the triplets by members of the nursing staff at the Karolinska Institute in Stockholm, Sweden. The triplets were also extensively interviewed about the presence of TS symptoms among family members.

The biological father and uncle were affected, as were the two male and female children of one of the female triplets. The triplets' paternal half-sister was affected with migraine headaches.

All members of this triplet set met DSM-III criteria for TS. The onset of TS occurred before five years of age, and at least one year following the placement of each triplet into separate adoptive families. The MZA twin pair displayed vocalizations that are persistent in one member. Vocalizations had also been observed in the male co-triplet (Segal et al., 1990b). One of the female triplets provided a childhood history of repetitive behavior suggesting the possibility of coexisting compulsive symptoms. The variability of obsessive-compulsive symptoms in TS has been documented (Price et al., 1985), and may explain the absence of these features in the other co-triplets. The sample studied by Price et al. (1985) contained two triplet sets, but the present report is the first detailed description of TS in a set of triplets reared apart. Clinical summaries for each of the triplets are provided in Segal et al. (1990a).

This case study of TS in an MZ/DZ triplet set, separated at birth, furnishes convincing evidence that genetic factors contribute to the onset and progression of this disorder. Its presence among first-degree relatives of the triplets, living both together and apart, further highlights the importance of genetic influences. Results from this case study, therefore, confirm findings from previous twin and family studies that have implicated genetic factors in the development of this disorder.

Sexual Life History Assessment

Many previous twin studies of sexual orientation have suffered from various sampling difficulties. There are, in addition, no systematically conducted studies of sexual orientation in twins reared apart, nor any studies that include female twins.

An early paper on sexual orientation in MZA twins included 2 male twin pairs and 4 female twin pairs, based on a series of 55 sets (Eckert et al., 1986). One pair of MZA male twin pairs was concordant, while the other MZA male twin pair was neither clearly concordant nor discordant. All 4 female twin pairs were clearly discordant. These data suggested that homosexuality may be more genetically influenced in males than in females, for whom this behavior may be an acquired trait.

Sexual life history information is currently available for 95 twin pairs (62 MZA and 33 DZA). Eight MZA twin pairs (3 male and 5 female) and 2 DZA twin pairs (1 male and 1 female), in which at least one member displayed homosexual or bisexual preferences, have been identified (Eckert et al., 1989). The single MZA male twin pair added to this series was clearly discordant for homosexual behavior. All female MZA twin pairs and the 2 DZA twin pairs were discordant. The homosexual females among the MZA twin pairs showed signs of delayed puberty and a larger maximum weight, relative to their co-twins. (One homosexual twin was lighter, but taller, than her co-twin at the time of assessment, due to a recent weight loss program.) These data, in conjunction with the earlier data, support the previous suggestion that male homosexuality may be due to a complex gene-environment interaction, whereas female homosexuality may be associated with environmental factors. Subsequent to this preliminary analysis, some additional cases of homosexuality have been identified among the reared-apart twins. Additional analyses of the various interviews and questionnaires are planned.

Homosexual behavior was observed among approximately 6.5% of the reared-apart twin individuals. Information on the population rate of homosexuality is reviewed by Pillard et al. (1981), and affords useful comparison with the findings from the present study. Predominant homosexuality has been observed among 4–5% of males and among 2% of females. The Kinsey studies, published in the 1940s and 1950s, indicated that the accumulated incidence of *any* homosexual experience was 33% for males by 25 years of age, and 20% for females by 40 years of age. Some participants were below these ages at the time of assessment. Present conclusions and interpretations, while provocative, remain tentative.

Future Directions

An overview of the psychiatric and medical assessment schedule used in the Minnesota Study of Twins Reared Apart has been provided. Neither data collection nor data analysis is complete at the present time. A key goal in the near future is the administration of the DIS to as many of the 42 twin pairs who were early participants in the study as possible. The availability of these data will enable more effective analyses of relationships between disorders. One such relationship is that between depression and anxiety, a topic of considerable current controversy (Kendler et al., 1987). Exploration of this issue has been delayed pending the availability of additional data.

Psychiatric and other life history data on the near-in-age siblings of reared-apart twins (i.e., the genetically unrelated children with whom the twins were raised) are now being collected. These siblings provide an alternative research strategy that is well suited to addressing the relative contributions of genetic and environmental influences on behavior. More specifically, it is possible to ask: are twins more similar to the unrelated individuals with whom they shared a rearing environment, or are twins more similar to their genetically identical co-twins from whom they were raised apart from birth? Finally, the availability of data on additional DZA twin pairs (which are less highly represented among the sample as a whole) will allow more stringent testing of genetic hypotheses.

The MZA design is also ideal for exploring gene-environment interactions that have been hypothesized to play a role in the genesis of disorders like substance abuse and antisocial behavior (Cadoret et al., 1983). We will be able to explore environmental factors in the rearing home, as measured by a variety of psychological inventories (e.g., Family Environment Scale; Moos & Moos, 1986), so that possible associations with psychiatric symptomatology can be examined.

The comprehensiveness of the research design will enable a number of explorations of relationships between psychiatric and nonpsychiatric data. EEG tracings (Stassen et al., 1988), personality data (Tellegen et al., 1988), and hand preference (Bouchard et al., 1986) are examples of the varieties of nonpsychiatric data that may be associated with psychiatric symptomatology. Previous studies have, for example, reported associations between psychopathology and handedness among twin populations (Boklage, 1977) that may suggest novel ways of examining the reared-apart twin data. It is, in conclusion, important to emphasize that the study of twins reared apart is one design out of several that are revealing with respect to genetic and environmental influences on behavior. Studies of twins reared together and studies of adopted individuals are critical supplements to psychiatric findings from reared-apart twin studies. This is because convergence among investigations employing varied methodologies will surely enhance confidence in the outcomes, and help to redirect future efforts along fruitful lines. The approach taken to both data collection and data analysis in this project has involved multiple procedures and techniques whenever possible.

Acknowledgments

This research has been supported by grants from the Pioneer Fund, the Seaver Institute, the University of Minnesota Graduate School, the Koch Charitable Foundation, the Spencer Foundation, the National Science Foundation (BNS-7926654), and the Harcourt Brace Jovanovich Publishing Co. Dr. Segal was supported, in part, by an award from the National Science Foundation, BNS-8709207.

We would like to thank the following people for the time and effort they have given to testing the twins: Margaret Keyes, Jeff McHenry, Elizabeth Rengel, Susan Resnick, Joy Fisher, Jan Englander, Ann Riggs, Dr. Kimerly J. Wilcox, Mary Moster, Dan Moloney and Ellen Rubin. We are indebted to our colleagues and collaborators on the Minnesota Study of Twins Reared Apart project, Prof. Auke Tellegen, Prof. David Lykken, Dr. Elke Eckert and Dr. Leonard Heston for their help and advice. We would also like to thank the Minneapolis Memorial Blood Bank, Dr. Herbert Polesky, Director, for the blood testing. Requests for reprints should be sent to Thomas J. Bouchard, Jr., Department of Psychology, Elliott Hall, University of Minnesota, 75 East River Road, Minneapolis, MN 55455.

References

American Psychiatry Association (1980) *Diagnostic and statistical manual of mental disorders*, 3rd ed. Washington, D.C.: APA Committee on Nomenclature and Statistics.

Boklage, C. E. (1977) Schizophrenia, brain asymmetry development, and twinning: Cellular relationship with etiological and possibly prognostic implications. *Biological Psychiatry* 12:19–35.

Borass, J. C., Messer, L. B. & Till, M. J. (1988) A genetic contribution to dental caries, occlusion, and morphology as demonstrated by twins reared apart. *Journal of Dental Research* 67:1150–1155.

Bouchard, T. J., Jr. (1983) Do environmental similarities explain the similarity in intelligence of identical twins reared apart? *Intelligence* 7:175–184.

Bouchard, T. J., Jr. (1984) Twins reared apart and together: What they tell us about human individuality. In *The Chemical and Biological Bases of Individuality*, ed. S. Fox, pp. 147–184. New York: Plenum.

Bouchard, T. J., Jr., Lykken, D. T., McGue, M., Segal, N. L. & Tellegen, A. (1990). Sources of human psychological differences: The Minnesota Study of Twins Reared Apart. *Science* 250:223–228.

Bouchard, T. J., Jr., Lykken, D. T., Segal, N. L., Eckert, E. D. & Heston, L. L. (1986) Anthropometric characteristics of reared apart twins: Fifty years of research. Presented at the 5th International Congress of Twin Studies, Amsterdam.

Cadoret, R. C., Cain, C. A. & Crowe, R. R. (1983) Evidence for gene-environment interaction in the development of adolescent antisocial behavior. *Behavior Genetics* 13:301–310.

Carey, G. (1988) Inference about genetic correlations. *Behavior Genetics* 18:329–338.

DePue, R. A., Slater, J. F., Wolfstetter-Kausch, H., Klein, D., Goplerud, E. & Farr, D. (1981) A behavioral paradigm for identifying persons at risk for bipolar depressive disorder: A conceptual framework and five validating studies. *Journal of Abnormal Psychology* (Monograph) 90:381–437.

Eckert, E. D., Bouchard, T. J., Jr., Bohlen, J. & Heston, L. L. (1986) Homosexuality in twins reared apart. *British Journal of Psychiatry* 148:421–425.

Eckert, E. D., Heston, L. L. & Bouchard, T. J., Jr. (1981) MZ Twins reared apart: Preliminary findings of psychiatric disturbances and traits. In *Twin Research 3: Part B. Intelligence, Personality, and Development*, ed. L. Gedda, P. Parisi & W. Nance, pp. 179–188. New York: Alan R. Liss.

Eckert, E. D., Heston, L. L., Segal, N. L., Grove, W., Lykken, D. & Bouchard, T. (1989) Homosexuality in MZ and DZ twins reared apart from infancy—developmental and personality variables. Presented at meetings of the Society of Biological Psychiatry, San Francisco.

Feighner, J. P., Robins, E., Guze, S. B., Woodruff, R. A., Jr., Winokur, G. & Munoz, R. (1972) Diagnostic criteria for use in psychiatric research. *Archives of General Psychiatry* 26:57–63.

Grove, W. M., Eckert, E. D., Heston, L., Bouchard, T. J., Jr., Segal, N. & Lykken, D. T. (1990) Heritability of substance abuse and antisocial behavior: A study of monozygotic twins reared apart. *Biological Psychiatry* 27:1293–1304.

Guze, S. B., Goodwin, D. W. & Crane, J. B. (1969) Criminality and psychiatric disorders. *Archives of General Psychiatry* 20:583–591.

Hankins, D., Drage, C., Zamel, N. & Kronberg, R. (1982) Pulmonary function in identical twins raised apart. *American Review of Respiratory Disorders* 125:119–121.

Hanson, B. R., Halberg, F., Tuna, N., Bouchard, T. J., Jr., Lykken, D. T. & Heston, L. L. (1984) Rhythmometry reveals heritability of circadian characteristics of heart rate of human twins reared apart. *Italian Journal of Cardiology* 29:267–282.

Hanson, B., Tuna, N., Bouchard, T. J., Jr., Heston, L., Eckert, E., Lykken, D. T., et al. (1989) Genetic factors in the electrocardiogram and heart rate: A study of twins reared apart and together. *Journal of Cardiology* 63:606–609.

Hathaway, S. R. & McKinley, J. C. (1943) *The MMPI*. New York: Psychological Corp.

Heston, L. L. & Shields, J. (1968) Homosexuality in twins. *Archives of General Psychiatry* 18:149–160.

Juel-Nielsen, N. (1965) Individual and environment: Monozygotic twins reared apart. *Acta Psychiatrica Neurologica Scandinavica*, Monograph Supplement #183.

Kendler, K. S., Heath, A. C., Martin, N. G. & Eaves, L. J. (1987) Symptoms of anxiety and symptoms of depression: Same genes, different environments? *Archives of General Psychiatry* 44:451–457.

Knobloch, W. H., Leavenworth, N. M., Bouchard, T. J., Jr., Eckert, E. D. (1985) Eye findings in twins reared apart. *Ophthalmic Paediatrics and Genetics* 5 (59):89–96.

Kohler, P. F., Rivera, V. J., Eckert, E. D., Bouchard, T. J., Jr. & Heston, L. L. (1985) Genetic regulation of immunoglobulin and specific antibody levels in twins reared apart. *Journal of Clinical Investigation* 75:883–888.

Langinvainio, H., Kaprio, J., Koskenvuo, M. & Lönnqvist, J. (1984) Finnish twins reared apart: III. Personality factors. *Acta Geneticae Medicae et Gemellologiae* 33:259–264.

Lykken, D. T. (1978) The diagnosis of zygosity in twins. *Behavior Genetics* 8:437–473.

Lykken, D. T. (1982) Research with twins: The concept of emergenesis. *Psychophysiology* 19:361–372.

Lykken, D. T., Geisser, S. & Tellegen, A. (1982) Heritability estimates from twin studies: The efficiency of the MZA design. (Unpublished manuscript.)

Lykken, D. T., Iacono, W. G., Haroian, K., McGue, M. & Bouchard, T. J., Jr. (1988) Habituation of the skin conductance response to strong stimuli: A twin study. *Psychophysiology* 25:4–15.

Moos, R. H. & Moos, B. S. (1986) *Manual: Family Environment Scale*. Palo Alto: Consulting Psychologists Press.

Newman, H. H., Freeman, F. N. & Holzinger, K. J. (1937) *Twins: A Study of Heredity and Environment*. Chicago: University of Chicago Press.

Pedersen, N. L., Plomin, R. & Friberg, L. (1985) Separated fraternal twins: Resemblance for cognitive abilities. *Behavior Genetics* 15:407–415.

Pillard, R. C., Poumadere, J. & Carretta, R. A. (1981) Is homosexuality familial? A review, some data, and a suggestion. *Archives of Sexual Behavior* 10:465–475.

Plomin, R., Pedersen, N. L., McClearn, G. E., Nesselrode, J. R. & Bergeman, C. S. (1988) EAS temperaments during the last half of the life span: Twins reared apart and twins reared together. *Psychology and Aging* 3:43–50.

Price, R. A., Kidd, K. K., Cohen, D. J., Pauls, D. L. & Leckman, J. F. (1985) A twin study of Tourette syndrome. *Archives of General Psychiatry* 42:815–820.

Robins, L. N. (1966) *Deviant Children Grown Up*. Baltimore: Williams & Wilkins.

Robins, L. N., Helzer, J. E., Croughan, J. & Ratcliff, K. S. (1981) National Institute of Mental Health diagnostic interview schedule. *Archives of General Psychiatry* 38:381–389.

Segal, N. L., Dysken, M. W., Bouchard, T. J., Jr., Pedersen, N. L., Eckert, E. D. & Heston, L. L. (1990a) Tourette's disorder in reared-apart triplets: Genetics and environmental influences. *American Journal of Psychiatry* 147:196–199.

Segal, N. L., Dysken, M. W., Bouchard, T. J., Jr., Pedersen, N. L., Eckert, E. D. & Heston, L. L. (1990b) Diagnosis of Tourette's disorder (Letter to the Editor). *American Journal of Psychiatry* 147:1386.

Shields, J. (1962) *Monozygotic Twins, Brought Up Apart and Brought Up Together*. London: Oxford University Press.

Sims, V. M. (1931) The influence of blood relationship and common environment on measured intelligence. *Journal of Educational Psychology* 22:56–65.

Spitzer, R. L., Endicott, J. & Robins, E. (1978) Research diagnostic criteria: Rationale and reliability. *Archives of General Psychiatry* 35:773–782.

Stassen, H. H., Lykken, D. T. & Bomben, G. (1988) The within pairs EEG similarity of twins reared apart. *European Archives of Psychiatry and Neurological Science* 237:244–252.

Tellegen, A., Lykken, D. T., Bouchard, T. J., Jr., Wilcox, K. J., Segal, N. L. & Rich, S. (1988) Personality similarity in twins reared apart and together. *Journal of Personality and Social Psychology* 54:1031–1039.

About the Contributors

MING T. TSUANG is Professor of Psychiatry, Harvard University; Director of Psychiatric Epidemiology, Harvard Schools of Medicine and Public Health; and Chief of Psychiatry, Brockton/West Roxbury VA Medical Center. Born in Taiwan, Tsuang received his M.D. from the National Taiwan University, and his Ph.D. in Psychiatry and Doctor of Science in Psychiatric Epidemiology and Genetics from the University of London. Author of more than 200 articles, chapters, and books, Tsuang is one of the foremost researchers in the field of psychiatric epidemiology and genetics and the major psychoses. Tsuang has received honorary degrees from Brown University and Harvard University, and is the recipient of numerous awards and honors, including the Josiah Macy Faculty Scholar Award at the University of Oxford, England; First Prize Clinical Research Award, from The American Academy of Clinical Psychiatrists; Rema Lapouse Award for Mental Health Epidemiology, presented by the Mental Health, Epidemiology, and Statistics sections of the American Public Health Association; the NIMH Merit Award on Psychopathology and Heterogeneity of Schizophrenia; and the Stanley Dean Award for Basic Research in Schizophrenia, American College of Psychiatrists. He is currently Associate Editor of *Schizophrenia Bulletin*, and a member of the editorial boards of *European Archives of Psychiatry* and *Neurological Sciences*, *Schizophrenia Research: An International Multidisciplinary Journal*, and *Social Psychiatry and Psychiatric Epidemiology*. Professor Tsuang's current address is the Harvard Departments of Psychiatry and Epidemiology; Chief, Psychiatry Service (116A), 940 Belmont Street, Brockton, MA 02401.

KENNETH S. KENDLER, Professor of Psychiatry and Human Genetics at the Medical College of Virginia was born in New York City and received his undergraduate education at the University of California at Santa Cruz, his medical training at Stanford University, and his psychiatric training at Yale University. His early research focus was in neurochemistry, but since 1983 he has worked exclusively in the area of psychiatric genetics. He is currently Principal Investigator on two large psychiatric genetic projects. The first is an epidemiologic family study and a linkage study of schizophrenia based in the West of Ireland. The second is a large study of a population-based sample of female same-sex twin pairs from the Virginia Twin Registry. He serves on the Epidemiologic and Services Research Review Committee of the NIMH and has been a member of the Task Forces on DSM-III-R and DSM-IV. He has published numerous papers in the area of his research interest, including the genetics of schizophrenia and affective illness and the statistical-genetic modeling of psychiatric disorders. Professor Kendler's current address is Department of Psychiatry, Box 710 MCV Station, Richmond, VA 23298-0710.

MICHAEL J. LYONS is Assistant Professor of Psychology at Boston University, Visiting Assistant Professor in the Section of Psychiatric Epidemiology and Genetics, Department of Psychiatry, Harvard Medical School, and Research Psychologist at the Brockton/West Roxbury VAMC. He was born in Long Island, New York. Lyons received a doctorate in clinical psychology from the University of Louisville, completed a clinical internship at Yale University, and a postdoctoral fellowship in psychiatric epidemiology at Columbia University. He is the co-principal investigator of the Harvard "Twin Study of Drug Abuse and Dependence," funded by the National Institute on Drug Abuse. Lyons has published numerous papers, and his areas of research interest include psychiatric genetics and epidemiology, schizophrenia research, and personality disorders. At Boston University he teaches undergraduate courses on abnormal and clinical psychology, as well as graduate courses on assessment, testing, and community psychology. He is also a practicing clinician. Professor Lyons's current address is the Department of Psychology, Boston University, 64 Cummington Street, Boston, MA 02215.

THOMAS J. BOUCHARD, JR., is Professor in the Department of Psychology at the University of Minnesota, 75 East River Road, Elliott Hall, Minneapolis, MN 55455.

PATRICIA A. BRENNAN is a doctoral student in psychology at the University of Southern California, Social Science Research Institute, University Park, Los Angeles, CA 90089.

REMI J. CADORET is Professor of Psychiatry at the University of Iowa College of Medicine. His address is the Department of

Psychiatry, Psychiatric Hospital, 500 Newton Road, Iowa City, IA 52242.

CHRISTINE A. CLIFFORD is Senior Lecturer in the Department of Psychiatry at the University of Tasmania, Australia.

RAYMOND R. CROWE is Professor of Psychiatry, the University of Iowa College of Medicine, Department of Psychiatry, Psychiatric Hospital, 500 Newton Road, Iowa City, IA 52242.

WILLIAM F. GABRIELLI, JR., is a researcher in the Department of Psychiatry, University of Kansas Medical Center, 398 Rainbow Boulevard, Kansas City, KS 66103.

WILLIAM M. GROVE is Assistant Professor in the Department of Psychiatry, University of Minnesota, Box 393 Mayo, Minneapolis, MN 55455.

ALISON M. MACDONALD is a research worker in the Genetics Section, Institute of Psychiatry, London SE5 8AF, England.

STEVEN MATTHYSSE is Associate Professor of Psychobiology, Harvard Medical School, and Associate Psychobiologist, McLean Hospital. His address is McLean Hospital, Mailman Research Center, 115 Mill Street, Belmont, MA 02178.

PETER McGUFFIN is Professor and Head of the Department of Psychological Medicine, University of Wales College of Medicine, Heath Park, Cardiff CF4 4XN, Wales.

SARNOFF A. MEDNICK is Professor of Psychology and Research Associate at the University of Southern California. His address is Social Science Research Institute, University Park, Los Angeles, CA 90089.

STEVEN O. MOLDIN is Postdoctoral Fellow in Quantitative Genetics, Jewish Hospital. His address is Washington University School of Medicine at Jewish Hospital, 216 South Kingshighway, St. Louis, MO 63110.

ROBIN M. MURRAY is Head of the Genetics Section and Professor of Psychological Medicine at King's College Hospital and the Institute of Psychiatry. His address is the Institute of Psychiatry, DeCrespigny Park, Denmark Hill, London SE5 8AF, England.

ROSALIND NEUMAN is Research Assistant Professor in Psychiatry. Her address is Washington University School of Medicine, 4940 Audubon Avenue, St. Louis, MO 63110.

MICHAEL O'DONOVAN is Medical Research Council Fellow in the Department of Psychological Medicine at the University of Wales College of Medicine, Heath Park, Cardiff CF4 4XN, Wales.

ANNE GERSONY PROVET is a doctoral student in clinical

psychology at Boston University, 64 Cummington Street, Boston, MA 02215.

JOHN P. RICE is Professor of Mathematics in Psychiatry at Washington University, School of Medicine. His address is Washington University School of Medicine, Department of Psychiatry, 4940 Audubon Avenue, St. Louis, MO 63110.

NEIL RISCH is Associate Professor of Public Health (Biostatistics) and Human Genetics, Yale University School of Medicine, Department of Epidemiology and Public Health, 60 College Street, P.O. Box 3333, New Haven, CT 06510.

RICHARD J. ROSE is Professor of Psychology and Medical Genetics. His address is Indiana University, Department of Psychology, Twin Studies/Laboratory for Behavioral Medicine, Psychology Building, Bloomington, IN 47405.

NANCY L. SEGAL is Research Associate in the Department of Psychology at the University of Minnesota, and Assistant Director, Minnesota Center for Twin and Adoption Research, University of Minnesota, Psychology Department, Elliott Hall, 75 East River Road, Minneapolis, MN 55455.

GEORGE WINOKUR is the Paul W. Penningroth Professor and Head of the Department of Psychiatry at the University of Iowa. His address is the University of Iowa College of Medicine, Department of Psychiatry, Psychiatric Hospital, 500 Newton Road, Iowa City, IA 52242.

Index

adoption method: adoptees' family method, 37; adoptees' study method, 37; and comorbidity, control and comparison groups for, 37–39; and etiology, 33–34; and foster children, 39; and genetic variability, 241; high risk method, 38; investigating biological families in, 41–44; and late adoptions, 38; measuring environmental factors with, 40–42, 247; measuring genetic factors with, 41–44, 247; measuring outcome in, 39; multivariate techniques, 36; MZ (identical) twins categorized by age-at-separation, 25–26; and natural children of adoptive parents, 39; and nosology, 33; partial separation design, 38; perinatal and postnatal factors, 40–41; and placement effects, 34–38, 43, 241–242, 252; prospective studies, 42–43; and psychiatric nosology, 33; rationale of, 34–38; registries and institutional records used in, 42–44; and SES, 22–23; vs. twin studies, 22–23. *See also* twins reared apart

adoption studies: bipolar disorder, 163; criminal behavior, 236–243; depression, 167–170; psychopathy, 238; schizophrenia, 124–125; sexual orientation, 255–256; substance abuse, 256–258; Tourette syndrome, 259–260

adoptive families: representativeness of, 40–41

affective disorder: diagnostic criteria, 153–155; differential diagnosis, 155; epidemiologic studies, 155–160; marker studies, candidate gene, 63–64; marker studies, classical, 54–56; marker studies, DNA, 61–62; marker studies, X-linkage, 62–63; morbid risk, 156; secular trends in, 105–111. *See also* bipolar disorder; depression; mania

age effect: and psychiatric illness rates, 94. *See also* secular trends

age/period/cohort effect. *See* secular trends

agoraphobia: family studies, 186; and panic disorder, 176; and phobic disorders, twin study, 186

alcohol abuse: cumulative risk and birth cohort, family study, 101; and depression, 167–170; lifetime prevalence, 103; in twins reared apart, 256–258

allelic heterogeneity, 84

Alzheimer's disease: segregation analysis, 87–88

Amish, the: and bipolar disorder, 8, 61, 160

anterospective studies, 17

antisocial behavior: adoption studies, 236–243; family studies, 231–232; twin studies, 231–235. *See also* criminality

antisocial personality: in twins reared apart, 256–258

anxiety disorders: family studies,

Genetic Issues in
Psychosocial Epidemiology

Series in Psychosocial Epidemiology
Volume 8
ANDREW E. SLABY, M.D., Ph.D., M.P.H.
General Editor